Dedication:

This book is dedicated to my Mom. Her love and support of everything I do have made me the strong woman I am today. To Richard for all the care, support, and help with my journeys in this book and in helping me write this book. You're a great friend Richard.

Running~Thru

A woman's solo thru-hike of the Appalachian Trail, done as training for the Triple Crown of 200's Races

By Stephanie White

2018 Appalachian Trail Thru-Hike and the Triple Crown of 200's all in one year.

Table of Contents

Chapter 1: The Appalachian Trail Adventure Begins

Trail lesson: *On the good days and bad days you still take one step at a time the same way. The good days will fade and the bad days will too, but just keep moving forward and enjoy the journey.*

This is the start of a new and big adventure! On February 18, 2018 I hiked the approach trail of the Appalachian Trail with my in-laws, then began my journey, solo all the way to Maine. I have officially just started my thru-hike on the Appalachian Trail. This is going to be an epic adventure that I've been looking forward to and preparing for a while now. Saturday night I was so excited that I could hardly sleep. I woke up throughout the night and couldn't wait for 7 o'clock to come around so I can get up and get ready to go.

My in-laws were nice enough to drive me to Springer Mountain yesterday. We drove up from Charlotte and stayed at the Springer Mountain Lodge. After checking in and visiting the Amicalola Waterfall we had a huge dinner. Then we went to bed early so that we could be prepared for today, Sunday. I enjoyed the hot shower for one of my last times that night and again the next morning before we set off. I'm going to miss hot water and showers but that is part of the journey.

Over the past week leading up to today, I had been quite anxious and nervous about the hike. Then it seemed to have disappeared and I was at peace on Friday before leaving. I know that this hike will be very difficult but will also be filled with moments of intense joy and amazingness. I look forward to growing as a person and finding a lot of answers out here.

The amount of love and support I have been shown for this adventure has been amazing. So many great people in my life have said kind words of support and love for me to do this. I was quite surprised actually. I know I am ambitious and have big goals but I did not expect how many people would understand and support me as much as they have. I'm very grateful for that.

As I hiked the trail, I kept a written dialog of my journey along the way with how it is going and what I discover mentally and physically. At first, I felt I would be doing all of this alone, but realize now how much support and love I have from so many people. Feeling alone is definitely not something I started this adventure feeling.

The three of us hiked up to the top of Spring Mountain. Once we made it to the top we had a nice lunch and I signed the book that states that I am starting this journey. I felt so official! While we were finishing up lunch, a guy came along with a cooler offering drinks to all the thru-hikers. It was my first experience with trail magic and it was very nice. Trail magic is considered an act of kindness given to hikers along the trail. This can be through food, rides into town, or anything that can help a hiker out in any way. The three of us ended up getting a bunch of pictures to thoroughly document this moment before I set off on my own. I was only about 100 steps away when I realized I didn't have my trekking poles and had to turn around to get them. It was funny actually. I'll have to get used to keeping those with me.

The reason I am out here in the first place is to get ready for some big races I have coming up this year. I am a long-distance runner and have completed 12 different 100-mile races over the past two years. I wanted to push those limits and felt this was the time in my life to tackle the 200-mile distance if I was ever going to. While doing research on which race to sign up for I came across the triple crown of 200's challenge. This is a series of three races, one month apart, on the west coast. They include the Big Foot 200 in August, the Tahoe 200 in September, and then the Moab 240 in October. After some thought, I decided to go for all three this year and that is why I am out here on the trail. Where I live in North Carolina, there are not many serious mountains to train on. I wanted to train my mind and body to be able to handle such grueling races. While the distance seems intimidating, the elevation gain and terrain are the biggest challenges I knew that if I was going to commit to these races I would have to take training for them seriously. After much research, I eventually was led to the Appalachian Trail. After some serious discussions with my husband, we decided I would take a year off working as a Paramedic and hike the Appalachian Trail, followed by the three races. I am essentially using the Appalachian Trail as training for the long-distance races shortly following my thru-hike.

Day 2

It has been rainy the entire day with a light drizzle, which hasn't been too bad, but it has not stopped the entire day. I didn't sleep well last night and could not seem to get comfortable. The reality is sinking in of the magnitude of this challenge. I had gone to bed a little early and was up by 6 am. I packed up my gear, which included wearing all of my rain gear. My pack is too heavy and I need to do something about that. I think it's mostly too much food so I'm looking forward to consuming a lot of it. A few miles ago I ran into a guy that had a whole tent set up and he was doing trail magic for all the hikers. He had a table with a whole bunch of food cooked for all of us passing through. This was the well know Trail Angel known as "Fresh Ground". He has been helping hikers for years along the trail and was a pleasant surprise to meet on the second day of my journey. There was already one hiker there, and I joined him as Fresh Ground warmed up spaghetti and meat sauce for me. He was so amazingly kind. He let us wash our hands and take fruit and Cliff bars before we left. I even had some Kool-Aid and a hotdog, neither of which I have had in many, many years. I took the hotdog to go since I was so full of the spaghetti. That will be my dinner. Amazing how my diet is going to be so drastically different than what I normally eat. The number of calories I'm blowing through is relatively high so I'm basically just planning on eating anything and everything I can, when I can. No major issues happening yet and I'm holding up well in this rainy weather. It's supposed to be rainy tomorrow as well with thunderstorms on Wednesday. I'm a little concerned about the thunderstorms and plan on staying in my tent if they're that bad, or a shelter. We will have to see how that goes. Other than that, everything is going great. It's wonderful to be out in nature and having the opportunity to be able to get up and hike this beautiful atmosphere every day!

Day 3 and 4:

I stayed at one of the shelters by myself for the night and it was quite an experience. I must admit, being alone way off the trail in the pitch-black dark was rather scary last night. Since I started this hike rather early in the hiking season, not many hikers are out on the trail yet. The shelter was full of mice and they would drop acorns off the shelf throughout the night, waking me up and scaring me to death. I was happy when the sun rose and I was able to get out of there. The rest of the day was a nice rainy hike and I made it to a shelter where there are a lot of friendlier hikers.

I slept great last night and I am ready for a 15.5-mile hike day today.

Last night was the best sleep I've had so far. Although it was very windy I was quite content in my tent. Things are going well so far and I'm enjoying this journey. There is lots of free time to think about life and ponder possibilities. I am appreciative for the time and opportunity to be out to experience this.

Chapter 2: Wet Noodle

Trail lesson: *The weight of items in your pack is determined more so by how much it all weighs soaking wet.*

Day 5. Feb 24, 2018

Last night I stayed at the camp area with two hikers that I had met the previous night. I was given the honor of giving one of the hikers a trail name. A trail name is given to a person on the trail by hikers that usually depict their character in some way. He was a very young guy right out of high school and he was an Eagle Scout. He was working on starting a fire with all of this wet damp wood and was able to get it going for a little while, and that's when the name "Young Eagle" came to me. I told him I wanted to offer him a trail name, told him what it was and he accepted it. He liked it a lot and I was happy to be the one to give it to him. The other hiker did not have a trail name yet and I'm waiting to see what comes to me for his name. A trail name is not something you can rush.

Last night was also a good lesson on how to deal with dampness and cold. After giving away my cheap sleeping bag that was soaking wet and weighed a ton, I was hoping that I could figure out a sleep system with what I had to work with. Not so much, I ended up sleeping in the shelter by myself while the other two hikers slept in their tents. I then had the opportunity to experiment with some of my gear. I used my sleeping bag liner, jacket, all my clothes, and my emergency Kevlar bag. I found out the hard way that once you get into the emergency bag, the body heat mixed with the moisture of the air quickly makes everything wet. Back to square one with my original problem, everything is soaking wet. I moved out of the bag and tried to sleep with my rain gear on, but was still cold. Finally, I had the idea to wrap the emergency bag around my legs and not get into the bag and that seems to help just enough so that I could get a little sleep. All the mice in the shelter were quite rowdy last night too, but once they started running around and playing with my gear I would just wiggle around in my emergency bag which makes a lot of noise and they would cut it out. I'm getting used to the fact that they are my sleeping companions and they don't seem to bother me anymore.

Day 6:

Today is going to be another new experience for me. I have decided to go into a town to resupply and charge my devices. I have plenty of food to keep going but I'm concerned about my sleeping system and my devices are almost depleted of power. By looking at the map there is a town in about 4 miles that I decided to go to. Since hikers don't have any means of transportation we have to resort to hitchhiking. I have never hitchhiked in my life and this will be a first. I'm afraid to do it, at least for the first time, so I've asked the other two male hikers if they are going into town and they both said yes. It is often difficult for males to get hitches so they try to hitchhike with a female when possible. I suggested that the three of us hitch a ride to town together. I would feel safer and they will have better luck getting a ride. They agreed and that is the plan for today. Both of them are much faster hikers than I am so I left a little early to try and get a head start. I'm still hoping to do 15.5 miles today so I don't plan on staying in town for too long. One of the hikers is going to stay with a friend of his while the other hiker, Young Eagle, is going to hitch back to the trail with me and we will both make it to the shelter that we planned on. He's younger but I do feel safe having him around. I'm going to take advantage of it while I can. I tend to sleep better when there are other people around for sure. Now if I can just figure out how to stay warm and dry I will be set.

Fortunately, the trails and mileage are going well. I'm going further each day than I had anticipated in the beginning. No major issues physically and I have more than enough food to feed on for a few days. My air mattress pad that I upgraded to before leaving is the one piece of gear that is working out absolutely perfect. It's very comfortable and warm. Today is one of the first days that it's not completely foggy and I'm able to see the trail and surrounding area well. It sure is nice to look around and see this beautiful trail.

Day 6 Continued

Hitchhiking was quite an experience. Two gentlemen hikers and I met up at the Road Crossing that we had all decided on. After we each had our picture taken hitchhiking for the first time in our lives, we set to work on trying to get someone to stop. We had already taken our packs off and had our thumbs out, the three of us. After that didn't work for about 15 minutes with multiple cars passing us, talk about getting used to rejection, I suggested that we put our packs on and hold our trekking polls so that people would recognize us as hikers. They agreed and we all put our packs back on. The very next car to pass pulled over! It was a woman headed into town to go to work. As we opened the door to thank her for stopping, she immediately said that she only stopped because she saw that I was a female, otherwise she would not have stopped, and she knew we were hikers. I had to giggle from my idea to hold our packs and trekking poles. She was very kind and drove us to a grocery store in town which was about 10 miles away. After stocking up on food, and charging our devices we set out to get back onto the trail. As we were leaving the grocery store, I spotted an antique shop and wanted to stop in to see if I could find a new blanket. I was cold last night and wanted to try a quilt. I walked in and immediately found a big blanket for $10. The woman said it was half off, and I said I'll take it. I will try it out tonight to see if it's any better. It's supposed to rain later on in the night so we'll see how that goes and how it holds up. Today has been the most beautiful day to hike yet. The temperatures are in the upper 60s and nothing but beautiful sunshine. I am not hiking as many miles as I wanted to because of the town stop, but that's okay. I was glad to restock on food and get my devices charged and anxiously awaiting to try out the new blanket.

Young Eagle and I eventually did make it back to the trail. It took us about 45 minutes and two different people had to give us a ride. The first gentleman had an appointment to go to and could only take us a certain distance, which we appreciated. The second couple to stop was a couple that lives in the area and often stops to help hikers out. We thanked all of them then were back on the trail. Young Eagle is a much faster hiker than I am so he went on ahead. I doubt I'll see him tonight because he was going to hike to a shelter further up the trail and I'm not going to make it there before nightfall so I will be at a campsite.

Day 7:

Last night was another frightening experience and I would like to explain why. I hiked into a campsite and set up my tent just as it was getting dark… made it to NC by the way, yay! There was one other hiker that was there, off in the distance, so I was happy to be with somebody and not completely alone. He was a good distance away, so we would not disturb each other. I tried out the new to me but used a blanket and it worked just ok, so I was happy about that. It even rained last night and I stayed somewhat warm, I was thrilled considering the past few nights. Around 5 o'clock in the morning, I was woken up to these odd noises. The noises sounded like a howling sound that seemed close by. It's hard to explain, but I could not really make out what it was and eventually concluded that it was a bear coming to get all the food in my pack and possibly slaughter me in the process. I lay wide-awake in my tent with my heart racing for 30 to 45 minutes. The noises became more frequent and louder as time went on. I had my taser of course and even zapped it a few times in an attempt to get whatever wild animal was out there away from me. That did not seem to faze it one bit and I became even more terrified! After about 45 minutes of staring at my tent and unzipping both exit points of my tent in case it came after me on one side, I could run out the other, I decided to do something about it. Since the noises were even more frequent and closer, it seemed, I was going to finally face my fate and see what the heck it was. My taser has a flashlight on it and I unzipped my tent and shined my light around the area to see that the guy in the tent off in the distance was rummaging around his tent with the light on and yawning loudly. No bear, wild cat, or Bigfoot! Whew!! I said are you alright and he replied that he was fine. He apologized if he woke me up.

After an incredible sense of relief that I was not going to be killed for the food in my pack, I zipped up my tent and laid back down. I just laid there for a few minutes and thought about how terrified I was when it was that guy yawning obnoxiously the whole time. I couldn't believe it! I went ahead and drifted off back to sleep for another hour and waited for the sunrise before I packed up my stuff to get hiking. As I was filtering some water that morning, Young Eagle came walking up to me. He camped in the same area but off over the hill so I didn't know he was even there. I had figured he'd hiked up way ahead of me but it was nice to see him. We ended up hiking together for the first 4 miles this morning before he decided to go a little faster today. It rained this morning but I was ready for it. I had all my dry things strategically packed up so it would not be affected this time.

I made it to the shelter this evening and had a good conversation with some of the more experienced hikers. They all advised me that I needed to upgrade my tent and sleeping bag, which I will do on Monday when I get into a town. I think the main problem is with condensation and that is getting my gear so wet. Apparently, cheap hiking gear won't cut it out here on the Appalachian Trail. The other hikers have informed me that the Smoky Mountains and the White Mountains can have some serious weather and I need to be ready for it. Since I'm freezing to death as it is now, I'm going to take their advice and get some better gear.

Chapter 3: New gear!

Trail lesson: *It is much easier to hike with fewer things. Much less to carry makes the hike more enjoyable and easier. I believe this could be applied to life as well. The less stuff we have the easier and more efficient life can be.*

Day 8. Feb 28, 2018

Today I hiked to the Rock Gap shelter, which was a little over 12 miles. Not a bad day and beautiful views. I arrived at the shelter early but did not want to push on to the next campsite or shelter because it would've been very dark by the time I arrived there, so I decided to take it easy and hang out at the shelter. I was all by myself and took the opportunity to air out some of my gear and collect a few things that I wanted to mail back when I made it to the town in Franklin. In about an hour before the sun went down, another hiker came by to stay at the same shelter and I was so relieved not to have to be alone. He gave me some advice about hiking which was useful and said he wanted to go into Franklin also, so we agreed to hitchhike together the next morning.

Day 9:

This was an unexpected day but turned out to be very interesting. That morning my fellow hiker and I hiked to Winding Stair Gap (3.7 miles) where we ran into Fresh Ground. Fresh Ground is a previous thru-hiker that sets up a tent with a ton of food and cooks for hiker's as they pass by all along the trail. He moves his set-up as the hiker's progress north on their journey. He is a really cool guy and is a former thru-hiker himself that comes out each year to do trail magic for all the hikers in the current year, and has been doing this for five years now. Really upbeat and positive person and it was fun to see him again. He loaded us up on pancakes before we went into Franklin. I had not intended to stay the night in Franklin but did want to get some errands done. I was able to get to an outfitter and purchase a new Big Agnes tent and good quality sleeping bag. Spending that amount of money on upgrading my gear was not in the plans but you have to adjust to what you have to do. The gear I had was not going to cut it out there but I was all set now and ready to get back on the trail. The errands took longer than expected and I ended up staying at the hostel. It was nice to get a shower and my clothes washed.

That evening I had a great time hanging out with all the other hikers that were at the hostel. I had met all of them along the trail and we all seem to end up at the same place at the same time that night. We all had a lot of food and enjoyed cracking jokes on each other and hanging out. I cannot remember the last time I had done that with a group of people and felt like I not only belonged to the group but that I was really in the moment and enjoying the time with them. It's so cool that we all feel like we are all Appalachian Trail hikers...not a paramedic, banker, teacher, and so on. We are all on the same level out here doing the same thing... hiking. I've never experienced this before. A typical hiker chat goes like this... what's your trail name, how far did you hike today, what shelters or campsites do you like best, how is your gear working out, what are you eating and what is best to pack out? And for the men, they discuss their next opportunity to get beer and pizza. There has been no discussion of professions, politics, celebrities, drama, or anything of the sort. It's all about the trail and how everybody is learning from it. It's a very important insightful journey for each one of us in very different ways and it's fun to hear about each other's perspectives on it.

They couldn't believe that I had a taser, so I had to show it to them. They couldn't believe how powerful it was and it became the subject of a lot of jokes that evening. All in all, it was a great time hanging out with hikers and "shooting the shit."

Day 10:

After a relaxing evening and a good night's rest in an actual bed, it was a great start to the morning. Young Eagle and I arranged a ride back to Winding Stairs Gap. When we arrived, Fresh Ground was up and cooking away as usual. We had another big pancake breakfast before we set out on the trail for the day. It was a little bit of a late start but made a great start to the day. It's a gorgeous day with the nice cool breeze so I'm hoping to cover some mileage today, and of course, I'm really anxious to try out my new tent and sleeping bag tonight. Have to enjoy the weather while I can because the next two days are supposed to be really rainy and cold. I'm not worried about it now since I have some better gear.

Next day: The new gear worked out perfectly. It rained last night and I just woke up dry. It's in the upper 30's and I'm warm, yay! The upgraded gear is great so far. It's supposed to rain all day today as well as tonight so we will see for sure how it all holds up. This was the warmest sleep yet, and I'm thrilled!

Chapter 4: The Great Smoky Mountains

Trail lesson: *There will be moments when you are exhausted, wet, cold, or really hungry, but embrace the feelings and know that it won't last forever. Just like with difficult situations in life, it will pass as well.*

Day 11. March 7, 2018

Today was a drizzly, rainy day but my longest hiking day so far. I was able to hike right through the Nantahala Outdoor Center, stopped for lunch, then packed out a big pizza. I hiked 18.6 miles that day. I hiked into the night and was able to use my headlamp for the last hour. When I arrived at the Sassafras Shelter, I was drenched. The shelter was full but my friend, Young Eagle was there and made room for me to squeeze in near him. I set up my things and changed into some warm, dry clothes before heading to bed. I was even too tired to have dinner. It was nice to sleep in the shelter because it POURED down rain really hard for most of the night. My sleeping bag was so worth getting, and I am so glad that I purchased a good one for this journey. I'll never take sleeping warm for granted again!

Day 12:

Today was a tough day with the weather. It poured down rain for most of the morning, and I was able to hike right through it. I feel like I'm getting better with this weather, slowly but surely. Another hiker had to go into town because he suffered from hypothermia yesterday. The conditions can be serious out here.

I was given a trail name today and now go by "Ultra Runner". People chose this name for me since I'm training for the 200-mile races out here and am leaving the trail for a few days in April to run an ultra-race (Salton Sea 81 miles). Plus, it pretty much sums up one of the reasons why I'm out here.

Tonight, at the shelter I learned how to hang a bear bag. One of the other, more experienced hikers showed me how to properly hang a bag, which I will have to do in the Smokies soon. The bear situation sounds serious in that area. Hikers have been attacked in the Smokies by bears wanting food, so hanging food is important and required. The bears are coming out of hibernation now and are in search of food.

Day 13.

Today was an easier day, Young Eagle and I hiked 7.1 miles to the Fontana Dam visitor center. From there we took a shuttle into the town of Fontana, which is the smallest town in North Carolina. One of the other hikers had already purchased a hotel room at the main hotel and allowed us to share it with him. For $20 I was able to sleep on my mat on the floor and get a shower. It was warm, dry, and comfortable, which I needed to rest up before hitting the Smoky Mountains the next day. While in town I did my laundry at the Laundromat and restocked my supplies at the general store down the road from the hotel.

A bunch of other hikers were at the same hotel as well and had another room a few doors down on the same floor. Between the two rooms, we all hung out together all day. We even went and had dinner together as a big group which was a lot of fun.

Day 14.

After breakfast, I arranged a ride back to the Fontana Dam visitor center and set off to tackle the Smoky Mountains. The first 5 miles are a climb up the first of four peaks that we will do. I paced myself and took it easy but knocked it out well. I did not expect to go as far as I did but managed to hike 17.1 miles today. That included taking a side trail to a fire tower, which I climbed and had the most amazing view of the mountains. It's kind of cool because at breakfast we were looking out the window and saw the fire tower on top of one of the mountains and were all discussing how we were going to hike to it that day. It was neat to be on top of that fire tower and look down at the hotel and see the distance we had come. Another hiker was at the tower as well named "Skittles" and he took my picture from the top.

The Smoky Mountains have one of the largest populations of bears, so all hikers are required to stay in a shelter. Section hikers have priority of staying in the shelter and if the shelter is full, thru-hikers have to pitch their tent and stay close by the shelter. After arriving at the Silers Bald Shelter it was full of section hikers partying and drinking. Noresta, Skittles, and I volunteered to pitch our tents and stay nearby. I didn't mind because I wanted a more peaceful and quieter place to sleep since they were partying so loudly. I didn't mind it though, as it's all part of the fun. It did get a little chilly last night though, in the upper 20s but the wind was really gusty. I woke up throughout the night because I was a little chilly but not terrible. Still, love my new tent and sleeping bag. I had packed a couple of extra layers of clothing because I knew the Smoky Mountains would be cold.

I'm planning on sending some winter gear home as soon as I get through this section, which will lighten my load some.

Day 15.

Today a few of us have decided to have a long hiking day. It's supposed to be very nice weather today and a storm is on its way so the weather will be rainy soon. We're hoping to cover as many miles as we can today while it's nice. I woke up at 5:30 am and hit the trail by 6:15 am this morning. It was absolutely breathtakingly beautiful seeing the mountains as the sun rose this morning. This has been by far the most beautiful hiking I've done. I took a bunch of pictures as the sun woke up the mountains. It was cool seeing a deer this morning also. Its hair was a little longer and thicker so it can survive the conditions up here. I saw two of them actually. Nice way to start the day.

All the other hikers think I'm tough because I don't have a cooking system. They have watched me eat cold mashed potatoes and tuna with my cold instant coffee and think that's pretty tough. Honestly, it doesn't bother me at all not to have warm food or coffee. I'm used to my food being cold and don't mind it at all.

I ended up stopping at 19.2 miles today because it was getting cold and dark quickly. I was passing a shelter and decided to stay there rather than push on the next 7 miles to the next one. There is a lot of ice on the trail and I already saw one hiker slip and fall. Two hikers fell yesterday actually, one of them being Young Eagle. He is now my "Wounded Eagle" buddy. He ended up coming to the same shelter as I did last night as well and took a zero-day the next day.

Luckily, the shelter had some space for us thru-hikers. A big group of section hikers arrived first, but there was plenty of room for us to stay inside too. That was nice and I was warm and dry last night.

Day 16.

I woke up early to get started on today's hike at 6am. I'm going to do 20.6 miles to the shelter and stay the night. I would like to take a moment now and say that I feel so fortunate to be out here on this gorgeous trail, early in the morning as the sun is coming up. There are so many beautiful trees, frozen streams, and birds chirping. It is so incredibly peaceful out here that it's hard to describe all of this accurately. I can't believe I am walking through the Smoky Mountain Forest right now and I am in the middle of all this beauty. The Smoky Mountains do have some difficult trails but on the other hand, some of the most gorgeous views and trails yet! Makes it all worth it. I feel so calm and at peace right now, since it is so incredibly peaceful out here. A level of peace I have not felt in a very long time.

I guess I'll get back to hiking so I can get as many miles in as I can before the bad weather hits. Supposed to be raining later on and then turn into ice this evening. Luckily, I have the extra warm clothes to get me through this section of the trail. I'm prepared now for it though. That and my sleeping bag is awesome. It's a big Agnes 15° bag. It's well worth it to spend more money on good quality stuff that will hold up in such harsh weather conditions out here on the AT. I learned that the hard way. But on the bright side, I can truly appreciate having warm feet at night so much.

Day 17.

Wow; the snoring! That is one big adjustment I've had to make on this adventure. When you sleep in a shelter there are bound to be people that snore. Last night this larger man was snoring so loudly it was crazy. I have found that when you're tired enough you eventually do get to sleep. Even with earplugs and a beanie on, curled up in my sleeping bag, the snoring is obnoxiously loud and clear. I'm getting used to it though. Sometimes you have to weigh your options and see if the snoring is worth being warm and protected by the shelter.

Today's hike started out rainy which was not an issue because I have plenty of dry warm clothes. I had planned to hike to the Davenport gap shelter but the weather turned out to be beautiful and I knocked out the 15 miles to get to the shelter by 3 o'clock, so I decided to keep going. There is a hostel in a little over 3 miles and I'm going to stay there tonight. It will be nice to get some solid rest, a shower, and a nice hot meal hopefully before I continue on. That will be my plan for today.

So, today is my third and final day in the Smoky Mountains. It has been an absolute pleasure hiking in these stunning mountains. Some of the most incredible views I've ever seen in my life up to this point. Also, the people have been very cool since it is spring break around here and a ton of college students are all out hiking. All the young energy is fun to see and feel.

I am starting to get brave. I have been trying to hike with people that I know. More accurately, meet up at shelters with people that I know. I have found that the plan doesn't always work out because people's plans changed by the hour, including mine. I never know how I'm going to feel during a 10 to 12-hour hiking day. I am not quite as scared being out here alone anymore. I have done about 95% of my hiking by myself, which I prefer. I'm not as scared of all the wildlife out here as I used to be either. I'm also getting better at hiking in the dark. I have a good headlamp and it is not so bad, quite peaceful actually. I'm also realizing that I will meet new people at each of the shelters or stay by myself which is not that bad at all. I don't feel the need to push my mileage or adjust my plan to stay with people that I know. Part of the experience is meeting new people along the way anyway. I must admit that today's higher mileage day is driven more by food than anything. While cold mashed potatoes and tuna packet is quite delicious, said sarcastically, I could go for some restaurant food right about now. I'm not quite famished but I am starting to daydream of a nice warm meal.

I hate that the Smoky Mountains portion of the Appalachian Trail adventure is almost over. On the other hand, I'm quite proud of myself for getting through these tough miles and high elevation in three days. I had packed enough food for six days; most people do the Smoky Mountains in about five days. I thought the weather would be worse than it was so I lucked out on that.

Later that day:
I was walking along this beautiful waterfall and stream when all of a sudden, I saw what I thought was a mirage. I saw Fresh Grounds tents and tarps and thought how amazing it would be to have some cooked food by him right now. I continued walking until people sitting down at the tarps started waving at me and I realized it was real. Fresh Ground WAS there and I went over and had the most amazing hot meal ever. I devoured two sloppy Joe's, one grilled cheese sandwich, an orange, and a small bag of chips. The food was so amazingly good and at just the right time. This is true trail magic. A bunch of my other hiker friends were

there and after we all ate so much that we were about to curl up and fall asleep, we hiked to the nearby Standing Bear Hostel. After a nice warm shower, and hanging out with the other hikers for a little while, I texted with Matt some before heading to bed. It was nice to stay indoors because it snowed quite a bit that night. I look forward to texting with him at the end of the day and seeing how he is doing. The Garmin satellite device comes in very handy.

Day 18.

Today is March 7th, which is Matt and my six-year anniversary. After not having cell service for the past five days I turned my phone on to check to see if I had any service and I did! I called Matt and had a great conversation with him. That has made my day and I am hiking with a big smile on my face. I love him so much and it was so great to talk to him, especially today. Each year has truly turned out to be better for us and I could not be more thankful for the love that he and I share together.

This morning I went to visit Fresh Ground and loaded up on some amazing banana pancakes before hitting the trail this morning. We have some high elevation to climb today so I needed as much fuel as I could get before setting off. It snowed a good bit last night and as I'm climbing higher in the mountains it is thicker and thicker snow. It's really beautiful. Although it's causing me to hike a lot slower, I don't mind. The beauty of it all worth taking it slow and enjoying it.

Chapter 5: Snow Everywhere!

Trail lesson: *Everything that happens in our lives, the good and bad things, are opportunities to learn and grow. Everything happens for a reason and we are given the opportunities to develop every day. It's our choice to learn from them or not.*

Day 19. March 11, 2018

Yesterday and today have been the most difficult hiking days yet! Hiking through snow that comes up to my knees at times, has been slow and difficult. The trail climbed quite higher after I made it out of the Smoky Mountains which produced a lot of cold weather in the higher altitude. It was a snow and ice storm that I have been hiking through. Luckily, I had plenty of warm clothes to get through it, but the hiking part was less than easy. Since the weather is so unpredictable and difficult, I pushed through it and made it through this section of the trail in two days rather than the 3 like I had planned on. That meant doing 15 miles one day and 18 the next over tough terrain. I am almost at Hot Springs now where I will stay at the Laughing Heart Hostel for the night. There is supposed to be even more snow so it will be nice to be inside tonight. Where did this snow storm come from in March?!?!

Last night I stayed at the Rocky Fork shelter with Skittles, Young Eagle, Lost Boy, and two other section hikers. The two section hikers were an older gentleman and between the two of them, it was the worst snoring I've ever heard in my life! To say I did not get any sleep last night, I mean that I maybe had an hour and a half of sleep in five-minute increments, if that. It was the most obnoxious and loud snoring I've ever heard. I was so happy to get up and get out of that shelter early in the morning. I had an early start since I was awake anyway, and tough 18-mile hike into town. I'm looking forward to getting into Hot Springs in about an hour and getting a shower and doing some laundry. But the most important thing is getting a good meal at the diner I've been hearing is so good. I slipped on the ice today but am alright, thank goodness.

The guys and I were discussing last night, at the shelter, how if people did not make it to a shelter before it became too dark, it could possibly have turned into a survival situation. The temperatures dropped to the teens, but the wind was crazy. The wind and snow almost blow you over at certain points of the trail. While it's beautiful to see the views, you have to concentrate on standing upright. The section hikers were telling us that in 2013 a guy froze to death in the Smoky Mountains and he was in a shelter. These are serious conditions up here and the weather can turn quickly. I am paying attention to the weather conditions, and I'm trying to be as cautious as possible. I leave with plenty of time to get to a shelter before nightfall so that I can get warm in my sleeping bag and avoid getting lost. I can see how dangerous it can really get if you lose track of where you are. The Appalachian Trail is so well marked with white blazes on trees, but with this huge amount of snow, you cannot see the trail or the markings in some places at all. Forget trying to figure it out at night in the dark. Therefore, I have been careful and hiking these more dangerous sections of trail in the daylight.

This morning when I woke up my shoes were completely frozen. I had them sitting out next to my sleeping pad overnight and they were frozen solid along with my full Nalgene water bottle. The whole thing was a solid ice cube. I'm not sure exactly how cold it was last night, but that must've been cold for that whole water bottle to freeze. That was a challenge this morning trying to warm up my shoes just enough to cram my feet into them. I had to sit on them for a while and transfer my body heat onto them so that I could loosen them up just enough to get my foot in them. Once I started hiking, I warmed up quickly.

A few of the hikers that I have met along the way are about to start the Smoky Mountains. Because the weather has been so bad the last couple of days they are staying at a hotel till the worst of it passes before they go into the Smokies. The Smokies are particularly dangerous because the elevation is so high at some points and some of the hiking is pretty difficult. Lots of steep climbs and declines which are dangerous with ice and snow covering them. I was glad to hear that they were staying put at a hotel until safer conditions.

Well, the hiker hunger has kicked in and is in full force. With all this strenuous hiking each day I am blowing through calories quickly. When I get the chance, I try to eat as much as I can. Some healthy choices, and some definitely not so healthy choices. I need the calories though so it is nice to just eat whatever I want and whatever quantity I want. I must admit, today I am looking forward to visiting the diner and having a great big warm meal. Some good food is definitely a motivation to get through parts of the trail quicker if I have the choice. As much as I love seeing all the views on the trails, my stomach is talking and needs some attention.

Day 20.

It was so nice to get into Hot Springs last night! The Laughing Heart Hostel was perfect. I was able to sleep in a back room all by myself with no snoring, having the best sleep I have yet to get on the trail. I went to the Hot Springs diner which was down the road from the hostel and ate an amazing dinner too. Did my laundry as well and was all set to go this morning. It sure felt good to be well rested, well fed and caught up on some much-needed sleep. Most of the other hikers I have been hanging with for the last few days decided to take a zero day at Hot Springs today. I wanted to push on because I'm meeting Janice and Andre tonight after a 15-mile hike. I'm looking forward to seeing them and going out to dinner this evening.

On my way to the trail, a few other hikers that were also heading out and I passed another thru-hiker that was stealth camping. Stealth camping is when you find a camping spot that is not designated or around other campers. It's more on your own while doing your own thing, away from other people. He had markers out by his tent and was letting all the hikers sign his tent as they passed. We stopped for a moment and put our trail names on his tent which we all thought was fun, and then kept going.

I had the pleasure of meeting three other hikers while at the Laughing Heart hostel yesterday. A female thru-hiker name "Zoom Zoom", because she is quick up the mountains, and her two other hiking buddies "Hot Sauce" and "CPU" which stands for central processing unit (He is from the Netherlands). They all left this morning at the same time I did and it was nice chatting with them for a while. As we were hiking, Zoom Zoom and Hot Sauce went on ahead and CPU and I hiked together for 5 1/2 miles. He does a lot of hiking all over the world so it was fun talking about all his adventures to exotic places. He was also interested in my ultra-running adventures so that's always fun to talk about too.

I have a small confession to make. CPU was wearing some cologne that smelled amazing! I definitely picked up my pace to keep up with him just to enjoy how good he smelled for as long as I could. Yes, I even ran a bit behind him to keep up. After 5 1/2 miles I had to pee so bad, I couldn't hold it any longer and had to let him go on ahead. Hikers out here don't even use deodorant, much less cologne so that was a nice surprise to experience. Plus, I found him really interesting considering all the traveling and adventures he's done. Really made the 5.5 miles go by like it was nothing.

Today will be an easy 15-mile day, and then spending the night with Janice and Andre in a hotel, the Comfort Inn. They are going to the Biltmore in Asheville Saturday to visit a high school friend of André's and wanted to stop by to visit me on their way. Matt is coming to visit tomorrow as well and I can't wait to see him too. I did not know that they were coming this early in my adventure but it will be a nice visit. Plus, it's supposed to rain a good part of Saturday and I'm going to take a zero day to spend time with Matt and his parents so that works out.

Day 21.

Today is my first "Zero-day" since I started the AT on February 18. A zero-day is when you do not hike any miles at all on that day. While a "Nero day" is when you do a very low mileage day. Those are usually town days to resupply. Yesterday I made it to the meeting place where I was going to be picked up by Janice and Andre five minutes before they arrived. Talk about perfect timing. From there we went to the hotel and checked in and then went to dinner with Andre's high school buddy. We had a great dinner and good conversation. It was fun telling all of them all about the trail and my adventures so far. Last night I slept so well in a nice comfortable bed being completely warm and dry. I will never take that for granted again.

This morning we had breakfast at the Waffle House and continued talking about AT stories. Andre has hiked many sections of the AT, so it was fun to compare experiences. Matt had a few issues with work he had to finish up this morning but he is on his way now to come to visit. I'm looking forward to him getting here and spending time with him!

After breakfast, Janice and Andre went on to the Biltmore with his friend while I stayed in the room waiting for Matt to arrive. Since I had the time, I took the opportunity to organize all my gear and thoroughly clean it. I had Lysol wipes I used to disinfect everything. I also traded out some gear, mainly clothes. I still have some cold/rainy weather I will be hitting this week and I still have plenty of warm stuff but a few things that I did not use as much as I thought I would have to go. My pack is much lighter from when I started and that makes it so much easier to hike. Before I started the AT, I put together two big boxes of gear that people could bring if and when they came to visit me along the trail. In those boxes I had items to resupply with that I would need along with the extra gear I could use. Those boxes of gear and supplies came in handy throughout the AT. Anyone who came to visit me while I was on the trail brought the boxes so I could switch out gear or resupply with items I had all ready to go in them. I chose to do this rather than mail myself food and gear through the mail with drop boxes. An issue a lot of hikers had was that they would mail their boxes to themselves and then have to wait a day or two because they arrived at the location of their box early. Also, many times people would get tired of eating a particular item and then be stuck with it if they had mailed themselves an abundance of that particular thing, such as Snickers bars. After the first month or so of eating something so often, it's nice to change things up. That was another reason why I didn't mail myself drop boxes. Everyone has their own strategy so they should do what works for them.

Day 22.

Back to the trail! It was the best zero-day yesterday that I could have imagined. Spending quality time with Matt was wonderful. Thank you, Janice and Andre, for making this happen!

This morning I am meeting a coworker, Bryan, on the trail. He is an avid hiker and wanted to hike some of the trails with me. We are meeting at 9 am and planning on hiking 12 miles to a Shelter. The next day he will hike back to his car and I'll continue on. I'm looking forward to the company.

Chapter 6: Listen to your gut feeling

Trail lesson: *When you have love and support at home (my husband), you have incredible strength within you and it doesn't matter where you are.*

Day 23. March 14, 2018

I was not excited about leaving my family but knew I needed to get back on the trail. A coworker named Bryan came to hike with me for the day and we hiked 12.2 miles to Jerry's shelter. It was fun catching up with him while we hiked. He cooked us dinner and brought an Éclair pastry from Amilies (Charlotte pastry shop) for dessert. That was a nice surprise. In the morning he ended up hiking back to his car while I continued on the trail. As I hiked on that day the weather conditions became colder and windier. I ended up getting to a shelter before dark, which was my goal. Another section hiker ended up getting to the same shelter and we were both alone for the whole evening. He talked all evening and most of the night telling all of these stories, one after the other. I was a bit uncomfortable and on edge around him. He did not do anything disturbing, but I was getting a bad feeling about him. I made him aware of my taser just in case he had any ideas. I slept the whole night holding onto it ready to use it.

I did not sleep well that night at all because of this uncomfortable feeling about the section hiker that I could not shake. Plus, the wind was becoming more forceful throughout the night. Temperatures went down to 16° with 20 to 30 mile an hour wind gust. In an open, three-sided shelter, you are exposed to the wind. My feet never did get warm and I woke up all night unable to get completely warm or feel safe. The section hiker told me multiple times that the weather is going to get worse the next day and that I should let him take me to a nearby hostel until conditions improved. Again, very odd that he said this over and over. He could very well have been a very nice and helpful guy but my guard was up. I forgot to mention that he had asked to take my picture a few days back as I passed him at a shelter. I was with another hiker so I didn't think it was a problem then but being out there alone with him now kept me on edge.

Day 24.

After a very cold, windy, and not so well rested night I decided I was going into town to wait out the snowstorm for the next two days. My feet were so cold and my shoes were frozen solid, again. The section hiker continued talking and talking as he did that night prior. He encouraged me to go into a hostel and wait out the storm because the weather conditions can get awful out there. I had heard of other hikers getting off the mountain as well so I made the decision to do the same. He repeatedly offered to take me the 20 miles to Johnny's Hostel in Erwin. Again, I was getting a funny feeling about him I couldn't shake. He had cell service and was making multiple phone calls but would walk away from earshot of me and that put me on edge too. I could not hear to who and/or what he was saying. That morning he was taking his precious time getting ready to leave the shelter which was fine with me. My feet were hurting so bad because they were so cold and I needed to get moving ASAP to warm up anyway. I went on without him to hike to the parking area where people park to do section-hikes, Sam's Gap. He had mentioned he had a small truck that morning.

As I was hiking, I decided I was not going anywhere with that hiker. I decided to listen to my gut feelings and it was telling me not to trust him. I wanted to get to Sam's Gap area and try to get a ride out of there before he arrived at the parking area also. Just felt the need to get away, that feeling became urgent the more I thought about it.

Now, this is going to sound weird. Before I went on the trail I had somewhat of a dream. In the dream, I was being attacked by a person driving a small red pick-up truck. The dream spooked me so much that I decided that if I ever came across anyone in a small red pick-up truck I would not go anywhere near him or that vehicle. Just as I made it to the parking area at Sam's Gap I spotted a small red pick-up truck out of the four other cars that were sitting there. It was the only pick-up truck there. I literally stopped and gasped out loud as I noticed his small red pickup truck! It was the exact red pick-up truck that I saw in my dream. I definitely was not going anywhere near it or that hiker whatsoever. There was another runner there that had just finished running and was about to leave. I tapped on his window and asked if he could give me a ride in the direction of the hostel in Erwin. He was going the other direction towards Asheville. I explained to him that I was getting a bad vibe from a hiker and wanted to get out of the area immediately. He offered to take me down the road to a location where I could figure out where I wanted to go. I did not have cell service by the way.

Just then, an angel arrived! A trail angel is anyone that provides help to a hiker along the Appalachian Trail, or any trail really. A retired police officer named Eric in a minivan pulled up and got out to talk to me. He said he was passing by and wanted to know about my hike and how it was going since he drives by every so often to see if hikers are around and need anything. I talked to him briefly and asked him if he was going in the direction of Johnny's Hostel. He said he was and would be happy to take me. I was thrilled! I put my pack into his minivan and we left. I explained to him that I was very uncomfortable being there and was getting some very bad vibes from a section hiker. He saw how disturbed I was and even showed me his police ID before we drove away. He was trying to calm me down, which he did. In my heart, I believe that my guardian angel sent this kind man to save me from a situation I was not supposed to be in.

Eric could not have been nicer. His wife is a nurse and he is a retired police officer. He told me that about twice a month he comes up to that very spot to see if he can help out any thru-hikers. He takes them down the road to the local grocery store to resupply if they need it. He then offered for me to stay at his home with his wife for the next two days while the weather was bad. I thankfully accepted their offer and they have been the sweetest people ever. He took me to lunch and then when his wife made it home from work we all went to dinner. He wouldn't let me pay for a thing and was so kind! I did my laundry at their house and had a nice warm shower. We had a great conversation all day long as well. They are very sweet people that just wanted to help and they did so much. I couldn't thank them enough.

If I hadn't listened to my gut instincts, I believe that that could've been a bad situation. I knew I was getting bad vibes from that guy but when I saw his red pick-up truck I knew for a fact that this was not a good situation to be in. I just can't believe that an angel came to my rescue within five minutes of me walking up to that very spot. Everything happens for a reason, but I believe that my angels sent an angel to help rescue me from a dangerous situation. That may sound crazy but that is what I believe 100%.

Day 25.

Today I took a zero day because of the weather conditions. The temperatures in the mountains were in the teens with high winds. I plan on getting back onto the trail tomorrow morning when the conditions are much safer. The shelters do not provide much protection from the wind so it can get quite dangerous up there if you're not ready for that kind of weather. My gear and I are ok with cold weather, but the wind-chill is not something I'm quite ready or prepared for when the temperature is very low already. It has been a great zero day here. Very relaxing and safe, while Eric and his wife continue to be incredibly generous.

Before going on the Appalachian Trail, I had heard that people are kind to hikers when they are in need. I was hopeful that I would experience this along the way if the situation arises, but did not expect to experience it so soon and at such a high level. Two days ago, after I was rather spooked by the section hiker and wanted to get out of that situation, the opportunity presented the situation to need some help. Now, I have received a lot of support and love from many people along the journey so far which I'm so grateful for. This situation is different because Eric and his wife are complete strangers to me and literally saw me on the side of the road needing to get away from where I was, after talking to me a minute. They took me in, have paid for all my meals, made me so comfortable in their home, allowed me to stay in their spare bedroom, helped me with my gear issues that I was having with my trekking polls by taking me to REI so I could exchange them, and even took me to get a pedicure, which I was blown away to receive. I have not had a pedicure in so many years and I was deeply touched by the generosity (he insisted I get one). They are both just kind people in general and wanted to help me along my journey. It's amazing the level of humanity and kindness that exists there are in the world. I was fortunate enough to experience it firsthand and can't wait to pay all their generosity and kindness forward when I get the opportunity. Just needed to share this story and try to express how appreciative I am to them and all the trail angels out there.

I am going to bed early tonight so I can rest up and have a good hiking day tomorrow. It's supposed to be warmer but rainy. I can handle that as long as it's not freezing cold with crazy wind. Looking forward to getting back on the trail and have thoroughly enjoyed my two days off.

Chapter 7: "The Trail Will Provide"

Trail lesson: *Many times, I would be walking along with my eyes looking down and my head down, focusing on the path. I would do this until I realize that my neck was stiff and then would look to my left, then to my right. Once I started looking around me more often my neck was much less stiff and I would get to see and enjoy nature and the wildlife that surrounded me much more. I think I need to do this in my own life after the trail as well. It's easy to get tunnel vision on the day to day tasks and forget to take a moment to look around to notice the beauty of our life that surrounds us. I will make it a point to be more present in day to day life and look around so that I can miss much less of what is going on around me.*

Day 26. March 21, 2018

After two days of rest with a trail angel at his home with his wife, I am back on the trail. I spent the majority of the day reflecting on human kindness. He generously opened his home to me for the past two days and could not have been any nicer to me, not wanting anything in return, only to be nice. It was a pleasure meeting him and his wife and spending time with them. For two days I was able to rest, recuperate, and then hit the trails strong today. It was quite windy out but I was happy the temperatures were a little bit warmer. I had to trudge through lots of snow today so it was a long day, but a very good day.

I find it wonderful that people can be so giving and caring just from the generosity of their hearts. I like to think that I'm a kind and giving person as well, but experiencing what I had the last two days makes me want to step it up so much more. We all get so wrapped up in our busy lives but I am determined to make a conscious effort to take more time to be generous and kind to others. That is something that is easily not given much attention to when life gets so busy.

Since the weather conditions have been harsh the last two days, most of the hikers have gotten off the mountain to let the storms pass. It was fun to return back to the trail with so many people. A lot of positive energy out here! I was fortunate enough to see some gorgeous views today. I hiked to the top of this one mountain and was able to get a 360 view of surrounding mountains. I took many videos and pictures of it just so I can remember it all later. It was breathtakingly beautiful. I felt like I was on top of the world. I did take a moment to pause, give thanks to the opportunity to be able to be there and experience such beauty. Not everyone gets to do this and I am so grateful for the opportunity.

I would like to take a moment to talk about food. Food is on every thru-hikers mind a good portion of the day and I wanted to share some of my thoughts on food. Food is wonderful but as a thru-hiker, it becomes amazing. Doing this strenuous activity for 8 to 12 hours, and sometimes more, each day you burn up some serious calories. Normally, at home, I try to eat healthily and make good choices. I decided that while on the trail, especially with access to food is limited sometimes, that I would eat what I like when I wanted to, and boy have I done that!

I don't think that I have ever appreciated food as much as I did until this adventure. I'll explain what I mean. At home, it's easy to go to the grocery store or go out to eat and get exactly what you want when you feel like it. Out here that is not the case. You pack food in your pack that will not spoil for 3 to 6 days until you can resupply. That means a lot of instant mashed potatoes, tuna packets, and oatmeal. I also pack a whole bunch of Fig Bars and

Nut Butter Cliff bars to eat throughout the day. As a thru-hiker now, I have noticed that I really appreciate a warm meal more than I ever have before. Going out to eat with hiker friends or my family who have come to visit seems to be a lot more meaningful and appreciated. Having a home cooked meal from the Trail Angels is also meaningful and special. Even eating my Fig Bars after a tough climb is awesome.

I love all kinds of food from healthy spinach salads to the Twix candy bars that I seem to be addicted to lately. Everyone out here eats Snickers bars, but I've only eaten one. My candy of choice out here is Twix and I have eaten about 1/3 of my weight in them. That's another thing, I have been eating like a horse and have not gained a pound. I eat whatever is available and as much as I want and burn it all off out here without even thinking. That is one of the many perks of being a thru-hiker. I already know that when I return home after the trail that it will be a tough adjustment going back to actual portions and healthy choices more often. But, I will enjoy it while I can while I'm out here to the fullest!

Day 30.

"The trail will provide"

I have heard this saying a few times while being out here in the last month since I have become a thru-hiker. It took some time, but I do feel like I am part of this hiking community and learning so much each day. I have heard a few people from time to time say the phrase "the trail will provide." When I would hear this, I wouldn't exactly understand what they meant but today I get it. I believe I understand what they mean by that phrase and what it means to me. The trail does provide, it's not just physical things like water and shelters, it's emotional and mental health out here. We all hit walls in our life. I could choose to share only the positive and happy side of the trail that is just amazing all the time but that wouldn't be real. I wanted to go on this journey for so many reasons and part of those reasons is for self-discovery and to see who I am and where I'm going.

Please don't have a concern, I am completely fine and not for a second deterred from getting to the end of this adventure, but want to express a few emotions I'm currently feeling out here. Because that's what reality is, the good and the bad.

Over the past weekend, I had a friend come to visit and hike with me. A friend that I had a relationship with for a long time that I care deeply about. My husband and I are in a polyamorous relationship, which means we have an open marriage, and this friend was my ex-boyfriend. We hiked 17 difficult miles one day but the difficult part was some heavy discussion that was necessary to have. The trail provided time and opportunity for that to happen. While it is difficult to accept things that you cannot change or control in your life, I do believe everything happens for a reason. I may not always understand the reason at the time or agree with the path, but I do understand that at the end of the day it's all meant to be.

Yesterday, when I went on to continue my thru-hike on my own after his short visit, I had so much thought and emotion running through my head. Good, bad, difficult, relieved, everything. The trail provides time to take a big look at yourself and figure things out, whether you want to or not. I was feeling quite emotional yesterday and was planning on having an easy day of hiking to a shelter that wasn't too far away or strenuous of a hike. I arrived at that shelter and it was still early in the day and I had plenty of energy so I decided to keep going. A little while later I turned the corner and was surprised to see Fresh Ground. I didn't know he was going to be there and he was all set up with all this food as usual. A whole group of hikers were there and were so happy and welcoming as I walked up. I sat down and had some amazing food but even better conversation. The trail provided me an opportunity to help me feel better from the emotional struggle I had been going through that day with these positive people. As I sat there and ate, we laughed so much about so many silly hiker stories and things that have been happening. It helped me get into the moment of what I'm doing and stop worrying about the past and future. I needed that at that time so much.

After Fresh Ground filled all of our bellies with so much food we could hardly get out of our chairs, he convinced us all to camp out there with him. The hike we had in front of us was a 5-mile mountain to a shelter with rough conditions at the top, where it would be windy since it's all exposed. It wasn't that difficult to convince us all to stay since we had so much fun sitting around and talking. I set up my tent and crawled into my sleeping bag before going to bed when the trail provided again for me. Another hiker came up to my tent and said, "Hey, Ultra Runner", I said "yes?", and opened my tent. He nicely said that tomorrow's a new day and gave me a fist pump before he walked to his tent to go to bed. Such small acts of kindness can mean so much to somebody at just the right time. Thank you to the hiker for taking a second out of your day to say and do that for me.

I don't think he understood what my emotions were and he didn't need to. He felt the need to do that and it was so appreciated. Those words have been rolling around in my head since he said them, "today *is* a new day." Today is a new day to hike these amazingly beautiful trails, to get stronger physically with every step I take, and to emotionally sort out what needs to be sorted out and to release what needs to be released. I am fortunate for this opportunity. I am forever grateful for what the trail is providing.

Everyone out here has a story and is hiking for so many different reasons. I find it fascinating to hear what everyone's story is. Fresh Ground was kind enough to share some of his stories with us last night and I was touched by what he said as well. We all have a story in life and it is necessary for us to go through the ups and downs for us to evolve as humans. I'm doing my best out here to evolve as much as I can too. The trail is providing the opportunity and time to heal. I wasn't even aware of all the healing I had to do when I started!

Day 31.

I can't believe I've been out here for 31 days! I was such a chicken when I started and I feel so much stronger now. I look forward to my new experiences and everything I will see in the next 31 days and beyond. Last night I was able to stay at the neatest shelter. There is a barn that has been converted to a shelter for hikers along the AT in the beautiful mountains. It's called Overmountain Shelter. Some of the other hikers were highly recommending that place so I decided to stay there for the night. The view was absolutely spectacular since the barn sits right on the edge of the mountain and you look out over into the valleys. We had a great fire and other hikers were there with me so it was a fun conversation too. The barn had an upstairs and downstairs area. I chose to sleep upstairs because of the storm that was coming that night. Good thing that I did, it rained really hard but it was really nice to be inside and listen to the rain all night.

Another unique feature about this one particular shelter was the privy. For those that don't know, a privy is a kind of like an outhouse. It's a small wooden structure, usually with a big hole in the ground and a seat. Not a commode that is flushable. You just go to the bathroom (#2) in a big dirt hole, basically. This privy was unique though because it had three metal wall sides, was small and was open in the front to the most beautiful view ever. You can sit there and do your business while looking onto the gorgeous valleys below. We all got a kick out of the privy and took some pictures, not while doing our business for the record.

Today I'm only going to hike about 9 miles. I had not planned on staying at a hostel but this one particular hostel has extremely high reviews and is known to have the best breakfast on the AT. I am not trying to rush through this adventure, so I'm going to "stop and smell the roses," and stay at the hostel. The weather is supposed to be bad tonight so why not. I will try to do a high mileage day Wednesday or Thursday to make up some mileage since today will only be about nine miles. I'm not too worried about my mileage actually. I'm moving along, making decent time and covering some ground, but still have a lot of fun along the way and that was the whole plan. The plan is not to have a plan actually. Just to push it when I felt like it, and hold back and enjoy the moments when I want to, and that is exactly what I'm doing.

Chapter 8: "The Climb"

Trail lesson: *I don't learn very much when things are easy, but I learned a HECK of a lot when things are hard. Embrace the difficulty and appreciate the opportunities.*

Day 32. March 25, 2018

We just had our 4th snowstorm in March here on the Appalachian Trail. This has been some crazy weather this year! I walked through the first two snowstorms and it was quite difficult. Then was presented with the opportunity to stay at a hostel for the last two and I took it. I stayed at the Mountain Harbor Hostel and had a wonderful last two days. Many of the hikers have gotten off the trail and a good number of them came to that hostel, including Fresh Ground. We all had so much fun hanging out together, telling stories, eating a lot, and having fun discussions. The hostel owners were so kind to us and even cooked us a big dinner the last night we were there at no charge. They made this huge spaghetti dinner with chicken parmesan, salad, bread, and dessert with sweet tea. There were at least 10 of us there and we ate everything; as in no food storage containers needed; every crumb! Hikers are hungry and we could put away some food. After dinner, a section hiker pulled me to the side and said, "I am impressed with how much food you can eat." I giggled and said that I am burning through some serious calories out there on the trail. Earlier that day we all weighed ourselves on a scale that was at the hostel and I am at 121 pounds, having lost 4 pounds so far (in a month). Considering the high amount of food I have been eating, I was surprised that I've lost anything. One of the reasons is because my pack is the heaviest of all the hikers that were at the hostel. This morning my pack weighed just over 40 pounds, more than I probably need. I am still working on condensing some of my stuff and I am still packing out way too much food, but I am lowering it and learning as I go.

It was really fun to take a zero day and hang out with all the hikers as we watched the bad weather blow through. Lots of snow and heavy wind, and we were all grateful to be indoors and warm. I am surprised at all this crazy weather, in March no less. I believe warmer temperatures are around the corner and am looking forward to that.

Last night I went to sleep on my bunk bed listening to Fresh Ground tell funny stories about past and current hikers. That was so much fun. Fresh Ground is iconic out here because of all the food and support he provides to all of us. He refers to us as his children and has to take care of his children, and he sure does. He not only feeds us till we are about to explode, but he offers advice and a listening ear whenever necessary. Besides that, he is a really funny guy with some hilarious stories and a great sense of humor. One of my favorite Fresh Ground quotes so far, "I have ADHD and for your entertainment and pleasure, I am not medicated." Every time I see him I am extremely full, have laughed a lot, and have walked away with a good piece of advice in regards to the trail in one way or another. It's an honor to know him.

Today is a beautiful day though and I am planning on catching up on some miles. My goal is to do 26.6 miles to the Laurel Fork shelter today. It's chilly outside, in the 20s but there is no rain or snow and the sun is peeking out, so today is the day to push it. I don't mind the cold temperatures but the wind is what gets me. That is why I am taking advantage of this less windy day.

Day 33.

Doing another high mileage day today since the weather is so nice. It's in the 40s but there is no rain, snow or terrible wind. Trying to get some miles while the conditions are decent. Something just happened that I wanted to share, mainly so that I would remember. I am at mile 20 right now and about an hour ago I switched my Pandora radio station to listen to some different music. I kid you not, I started climbing this difficult incline of a mountain when the song by Miley Cyrus "The Climb" came on. An extremely therapeutic moment just happened, let me explain. I feel like I'm in a good state of mind and have been very happy the last couple of days, especially after so much rest and laughing with the other hikers lately. That song came on and I cranked it up. No other hikers were around, thank goodness, because what happened was YouTube worthy. The words had so much meaning to me at the moment and in my life. Out of the group of hikers I was hiking with today I was at the end, which is ok. Sometimes I just hike slower but I get where I need to go and ended up doing 30.2 miles total that day. (Listen to the song if you have a moment). I turned that song up and sang along to the words and even shed some tears. I felt my heart healing and my spirits just illuminating. I sang and cried up the mountain and just released so much tension that I didn't even realize I had! Frustration for trying to help people that won't listen, friends that had different intentions than I did and led me on, frustrating people at times, and so much more, just released! The radio station is on random so I didn't even know that song would have come on. I felt so good after it was over! It is moments like these that can be significant sometimes. Such a big sense of release and relief after that song was over that I had a spring to my step afterward. Well, to be honest, a spring in my step after I finished climbing the mountain and caught my breath since I was winded. My pack is still heavy and it's not easy climbing these big mountains, but afterward, I was skipping down the mountain with a sense of relief and joy. I may have terrified every wild animal out there in a half mile radius. Totally worth it though! So much of my day is focused on gratitude, and being so thankful for all that I have and can do. I guess I had forgotten to release the negativity I held inside. It felt so good to do it! There goes the trail providing again.

Day 34.

Yesterday was the longest hiking day so far. Another hiker and I ended up hiking 30.3 miles and getting into a shelter at 10:30 at night. We wanted to go as far as we could because the weather would be so bad today. At night we walked through cow pastures and the moon was so bright that we turned our headlamps off and walked the path in the moonlight, it was awesome. I woke up at 6:45 the next morning and was on the trail by 7:30 am to try and get into Damascus VA before the weather turned bad. Made it to Virginia today!

I had no idea until today that it could rain and snow at the same time. I was hoping to get into town before the rain started and had stupidly left my rain gear in the bottom of my pack. I did not want to take everything out and get everything wet, just to get it so I decided to push on faster to get to town. That meant that I was soaking wet and with the cold temperatures and wind, was just frozen by the time I made it to Damascus. I was starving too, I know big surprise! My instant mashed potatoes, beef jerky, and oatmeal are not cutting it for meals while I'm hiking so far, each day. Before I get to a town I am so hungry that I can hardly stand it. I intend to get a bigger cooking pot in town and start cooking some real pasta meals from now on. Some towns are right along the path of the AT, such as the town of Damascus.

I found the place we were staying called the Lazy Fox Inn. This place came highly recommended so I wanted to check it out. It was the cutest little Victorian house I have ever had the pleasure of staying at. The woman that owns the house is 93 years old, and she does everything to run it. She loves hikers and does everything she can to make us comfortable and feed us. After settling in and getting a hot shower I set out to get some food with big priority. I went to a really good restaurant called Mojo's and just devoured so much great food. It was great to get warm and some delicious food, while feeling like an entirely new person.

Made it back to the house and took a nice, long, hot bath which I needed desperately after these three long back to back hiking days. The first day I did 26.6 miles, then 30.2 miles, then 18 miles into town. My ankles and shins were so sore but the Epsom salt bath helped a lot. It was an old-fashioned bathtub and it was fun to soak in. It felt great to wash some clothes, eat and rest up in a dry and warm place. I found out later that two other hikers had gotten off the trail yesterday because of hypothermia. Once you get wet in these cold temperatures, it is difficult to warm up. A lot of hikers stayed in town until the weather improved. I was happy that I made it to town before it became worse later on in the day.

Day 35.

The owner of the Lazy Fox Inn, the 93-year-old woman and her 82-year-old friend made us a tremendous breakfast! At 8 o'clock in the morning, they rang a little bell and everyone came to the table to have breakfast. Two other hikers that I knew were there also and it was so much fun catching up with them (Humming Bird and Black Bird). They are a really cute, young couple that met each other on the trail. This older couple was the cutest couple I've ever seen. I hope to be half as adorable as they are when I grow old. They had the best time cooking and chatting in the kitchen that we all could not help but just watch and listen to them as they prepared breakfast. The Lazy Fox Inn was one of the best places I have ever spent the night. It was such a beautiful house that is so warm and friendly that the vibe in the home was just incredible. I can see this place being a great romantic getaway vacation sometime in the future. I really hope to come back here.

I was talking with the elderly woman last night when I noticed something on the back of her shirt, behind her right shoulder, and went to pick it off, thinking it was some lint or something. As I got closer, I noticed it was a pin of a little angel. She said it was her guardian angel and my heart melted. She had on a different shirt today but her guardian angel was with her again.

After breakfast, I did some errands, on foot because we are hikers after all, and I needed to get some food for resupply. After that, I found out that Fresh Ground was at a hostel nearby and was cooking everyone lunch. I made my way over there and had a great time chatting with all the other hikers and having a wonderful roast beef lunch. As you can tell, eating is a big priority for hikers along this AT journey. Damascus is an iconic trail town with so much history. A great place to visit and I definitely hope to come back one day. It's a bike town too so there are bikes and bike shops everywhere, along with hikers. Just a good vibe of a town. Trail Days is also held here each year in May. I may go next year and represent the class of 2018, not sure yet.

Chapter 9: "The Ponies"

Trail lesson: *There will be times when people want to start a conflict with you. This is the perfect opportunity to focus on how you want to handle the situation. It is hard for someone to argue with you if you do not argue back. While in such a situation recently, I chose and hope to continue to choose, to not respond with negativity and to send that person love instead. Most of the time the situation will be resolved on its own with time. No need to give your attention and focus to negativity when you don't have to.*

Day 36. April 7, 2018

After another wonderful night's sleep at the Lazy Fox Inn and a wonderful breakfast, it was time to leave Damascus. The cute elderly couple at the lazy fox made us a terrific breakfast again this morning. I checked to see if she was wearing her guardian angel pen and she sure was. I was able to get a picture with her this morning which was nice. She was so sweet and said that I need to come back and visit her so she can take me to a play in the next town over. If I get the chance to come back and visit while she is still alive, I definitely will.

I was not in a hurry to get out of town because I only have to hike 12 miles today to a cool destination of where I get to sleep under a big bridge along the Virginia Creeper Trail. So, I took my time packing up my pack and went to the post office to mail home some supplies that I didn't need to carry. Then went back to Mojos for lunch and packed out a sandwich to have for dinner tonight. I sure did have a good time in Damascus and really hope to return one day.

I am hiking along the trail next to a beautiful white water rapid along the Virginia Creeper Trail. I made up this portion of the AT a few days later since I wanted to see the Creeper Trail for the first time. The sound of the water is so soothing and calming. I had to turn off my music to just enjoy it. Plus, I'm walking on a path with no snow, it's not raining, it's not snowing, so it is a gorgeous beautiful day. I've had about enough snow for a long time so I am welcoming the sunshine with open arms.

Day 37.

Last night I slept under the Virginia creeper trail bridge next to a white-water river. It was incredible! Listening to the rushing water all night and having my whole tent illuminated by moonlight was so romantic. It's a place that most hikers walk right past but another hiker told me about the place and that's where I stealth camped. It was a fun spot and I hope to do it again one day.

I find myself thinking about a particular hiker that I met a few days ago. His trail name is "Loner Boner" and he is on his fourth thru-hike of the Appalachian Trail. That means he's completed three previous thru-hikes. I was able to meet him at a shelter one day and spoke with him briefly as I passed through, and saw him again up the trail a few days later (I zeroed and he caught up). He is a significant hiker because he is 77 years old and has full-blown bone cancer. He is out here taking this trail from shelter to shelter and having a great time. I spoke with other hikers that know him and they all laughed as they told me that staying in a shelter with him is a challenge because he talks to his wife on his satellite phone for 2 to 3 hours every night very loudly. I think that's so cute. I admire his enthusiasm for life. If I ever feel like I'm struggling with something, especially physically, I will remember "Loner Boner." He is out here doing it.

I have been looking forward to hiking this particular part of the trail today. Today I am going to enter Grayson Highlands and that is where the wild ponies are. I will be staying at a shelter that is right in their pasture area so I hope to see them and play with them a little bit. I've heard that they are very friendly and love hikers so we will see if they come to visit.

Day 38.

Yesterday I made it to Grayson Highlands and saw all the wild ponies. They were so cool! I saw 15 of them total, and one of the mothers had a tiny little baby with her. I took a bunch of pictures, and it was a very cool experience. So glad they were out and I was able to see them.

Last night the wind was crazy! I heard it was going to be quite windy that night so I decided to push onto the next shelter which was 5 miles further. Good thing I did because the winds were 40 to 50 miles an hour and it would sound like a train as they blew through the original shelter I had planned to stay at. And I heard there was a wild skunk living in the upper-level portion of that shelter that made the smell less than appealing in there. No thanks! Oh, yesterday I made it to the 500-mile mark, another milestone.

This morning I was up early to hike on to Fresh Ground who is set up 7.3 miles away in a public camping area. I'm looking forward to some banana pancakes so that had me moving early this morning.

Day 39.

Tomorrow will be an exciting day. I have my mom, Brandon, and the dogs coming to visit me tomorrow and I can't wait to see them. They are going to camp out and spend the weekend with me at the camping area where Fresh Ground is set up. It's the perfect location to meet and hang out for the weekend. Since I plan to spend some time with them when they arrive, I planned a big hiking mile day today. I arranged a ride from a guy that was there supporting his wife on her hike who tried to charge $30 dollars to drive 15 road miles out, further up the trail so I could hike back to that location the following day. It was a 15-mile car drive with a 33.1-mile hike through the trail back to the Fresh Ground set up a location where I was meeting my Mom. I gave him $20 and learned my lesson with him, to settle on a price before getting the ride. He was trying to make some money by being slick and springing that on me when we arrived at the location I would spend the night at, and then hike back to my starting point the next day.

Day 39 was my biggest hiking mile day with 33.1 miles. I was off to a good start, it is noon and I've already knocked out 12 miles. It's an amazing day because going in the opposite direction I have seen so many hiker friends. I did not realize that we were all so close in proximity to each other. As I'm going the opposite direction of the trail I have passed a ton of them. It's been so nice to catch up with them and see how they're doing, especially seeing some that I hadn't seen in weeks. I saw Nor'easter, Young Eagle, Hummingbird, Blackbird, French Fry, Weatherman, and a few more.

I was thinking about how shy I was in high school, with few friends growing up because of my shyness. No reason to go back to class reunions since I have nobody to go see. That applies to college as well. Over the years, I have found my voice and social abilities, better late than never. Seeing so many hikers today makes me feel very much a part of this hiking community. I took a picture the other day while we were in the town of Damascus that said "Welcome Hikers class of 2018." This is the first time I have felt like I belong to a class of any kind. It's really cool and I am proud to be part of the hiker class of 2018. It's a great group of hikers and I have made some lifelong friends already.

Later that day:

I arrived at the campground area just as the rain started so I was just in time. Fresh Ground had gone to bed, but as soon as he saw I and another hiker had arrived, he insisted on getting up, pulled out all his cookware, and making us an amazing dinner. He cooked us a great big hamburger, grilled cheese sandwich, and sliced up potato fries. We were so hungry and it was just so kind of him to get up and cook for us even though it was 9:45 pm. I was proud of myself for hiking 33.1 miles in one day, felt accomplished, though super sore and tired.

Day 40.

I slept in till 7:30 am when I was woken up by the smell of banana pancakes that Fresh Ground was cooking. I had three of them this morning and feel like I'm in a food coma right now. I'm so excited to see Millie and everyone today and can't wait for them to get here. I'm quite sore from the hike yesterday so today's day off will be very welcomed. I'm planning on just hanging out at the campsite all day today and setting up their camp supplies when they arrive.

Day 41.

I've had the best time visiting with my mom, Brandon, and the dogs. My mom brought my dog Millie to visit and it was so good to see her. I didn't want her to forget who I was and she hadn't. We stuck together the entire weekend and didn't leave each other side for a second. A love for a pet can be strong, Millie and I definitely have that strong bond. A few days ago, as I came out of Damascus, I hiked the Virginia creeper trail and had skipped 14 miles of the Appalachian Trail to do that. I wanted to go back and do that section so that I can complete the entire Appalachian Trail on this adventure, a "purist." Millie and I hiked it together and it was just like old times hiking together. I had such a good time running through the Virginia Mountains with her and took video and pictures of it which I'll look back on throughout my journey here. She even snuck into my tent when we made it back, crawled into my sleeping bag and took a nap. I found her in there and she was adorable.

My mom was wonderful and brought so much food to feed not only me but all the other hikers. We set up camp near Fresh Ground so it was a stopping point for all the hikers in the perfect location for us to spend this weekend visiting. She made trays and trays of macaroni and cheese, marinated chicken, cookies, banana bread, and a special chocolate cake for me. She brought a ton of other snacks and food for hikers to be able to easily take with them as they left camp. The day before they arrived I hiked 33.1 miles to get to that location and they arrived the following day on Friday. We were able to spend all day Friday, Saturday, Sunday, and then they dropped me off where I had left off in Atkins Virginia before they went on home. I hated to see them go but it was such a wonderful visit. My mom and Brandon were even given trail names. My mom was named "ultra-mom."

Brandon kept climbing these trees and even tricked a lady thinking she saw a bear so he became "Little Bear." One night all the hikers came together and made Brandon a hiking staff with his trail name "Little Bear" on it and gave it to him. It was really cool. Trail names are given to a hiker by other hikers after they have gotten to know that hiker. It's usually based on something you do out on the trail. The hiker can accept the name or decline it, in hopes to have a different one. Some hikers change their name as another name becomes more suitable for them.

Day 44.

The weather cannot make up its mind about what it wants to do. Today it has been sunny, then rainy, then really windy, and now raining again. It's been a roller coaster of weather today but that keeps it interesting. It's not freezing cold so it's all good with me. Today I am doing another big mile day and going to end at 28.6 miles. I'm hoping to make it to a place where I can get a shower tomorrow which will be a 22-mile day so that's why I'm pushing it today. Plus, it's supposed to rain all day tomorrow. This weather seems to be unpredictable so you never know. I haven't had a shower in a week and I would like to have one so I will push through whatever weather is thrown at me. Hot running water is such a luxury.

Today I'm feeling really good and uplifted. It was a nice visit with family and hiking has been really fun lately. I'm happy that I'm able to do some bigger mileage days without too much of a problem, just general soreness and that's about it. At night I prop my feet up over my pack and by morning I feel good to go. I feel like I have worked through and let go of some issues I had on my mind that were necessary to release. Feeling at peace and harmony with everyone and everything and that feels good. Seems to allow me to connect with nature out here a little better and focus on not only the trail but positivity in my life. I know I say this all the time of how grateful I am for everything, but I truly am. I have a very fortunate life filled with so many opportunities. I'm excited to see what will happen when I am finished with this trail. I do have lots of options and I'm excited to explore some of them. We will see what happens.

Chapter 10: McAfee Knob and the Cliffs

Trail lesson: *Observe and acknowledge your emotions, you're having them for a reason. Suppressing them can be dangerous. Look at emotions as guidance rather than a negative thing that should be suppressed.*

Day 50. April 12, 2018

This morning is the second morning in April that I woke up to my tent being covered in snow. I'm not complaining about it because it is beautiful. I'm staying nice and warm in my tent with my gear so I'm fortunate not to be too cold on these chilly nights. What a difference good gear makes! This morning I woke up before everyone else in the shelter and started to hike. I was the first one to make tracks on this beautiful snow and it's very beautiful out here. I never would've expected so much snow in April but these are the mountains of Virginia so I guess anything is possible.

The last few days have gone well. I've done some high mileage days and been sore so I'm trying to take it easy the next couple of days. Yesterday I did 25 miles and today I will do 18. Hiking in the snow is a little more strenuous, so I'm trying to be careful. I slipped and fell twice yesterday, and slipped and fell once already this morning. When I do fall I try to go with it so that I don't permanently hurt anything, I've been fortunate.

I've been thinking a lot lately about opportunities in my life. I do have lots of options and I'm looking forward to what I will be doing after I'm finished with the trail. I'm also getting excited about an ultra-marathon in California that I'm going to run at the end of this month. I will be getting off the trail for about a week to drive out there, or possibly fly, and do that race and then come straight back to the trail. I think it will be a nice break and I'm looking forward to it.

Day 51.

After a good night's sleep in my tent, I woke up and was on the trail before 8 am. I have 8 miles to go to get to Fresh Ground and his banana pancakes are calling my name. It's a beautiful morning with a lot of moisture in the air but no rain or snow, a nice change from the past couple days. There are a lot of birds out happily chirping around which is pleasant.

A few people have asked me some questions about the trail so I thought I would write a few of them down. Since I've been on the trail for 51 days now, here are a few statistics. I've taken 6 Zero days (no hiking days), mainly because of the weather. I had to sit out some snow storms on this adventure as you all know. I'm currently at mile 690. Most of the time, I typically hike to a shelter at night. I prefer staying around shelters because usually there are more people, there is a water source close by, and there is a privy (bathroom hole) if needed. I usually sleep in my tent since there are people that snore really loud in the shelters and I prefer peace and quiet. Now that I have a good tent and sleeping bag it is way more comfortable using that. Although, if it's raining really hard or snowing badly I will sleep in the shelters.

People have asked me if I am afraid out here. The first week or so I was afraid a lot, now not so much. I'm getting the hang of trail life and how it works. I have had a few instances where I have been cautious with some people but I just try to avoid the situation. Luckily that's not happened too often. As far as animals and nature, I'm not worried at all. I have done some night hiking by myself and it's actually rather peaceful. If I do get a little concerned with animals around I will just play my music out loud rather than through my headphones, which I typically do anyway since I hike by myself. I am more concerned with walking along the highway into a town than I am about walking alone in the dark in the woods, by far. Being out in nature is almost majestic and I appreciate the peace and serenity it has to offer. A lot of times I hike without any music or anything so I can just enjoy this peace.

The other day at camp somebody asked me what I listen to while I'm hiking. I have a variety of things that I listen to. I have a lot of audiobooks-mostly on spirituality and self-development stuff, podcasts, music, and recently signed up for Google Play. Pandora does not work very well off-line but Google play does and I'm enjoying that music a lot. I would say that I mostly listen to audiobooks. I have some good ones I have listened to a few times since I've been out here. The only electronic device I carried on the trail was my phone and the satellite communication device, the Garmin InReach.

People have asked if I have a "tramily" aka Trail family. A tramily is a small or large group of hikers who stick together the whole way. They prefer sticking with the group for the companionship and safety of it. I do not have one and prefer to be solo. I've met some wonderful hikers and people along the trail but I am a solo hiker. I like to do things on my own schedule and time and that includes my hiking days. Some days I like to go further and some days I like to take it easy. I never know when those days are so it's nice not to be obligated to a group of people that need to go at a certain pace or distance.

As far as food and what I carry, it varies. I will look at the map and carry enough food usually for 4 to 7 days. I like to change my food selection up so I don't get bored with it. I've been really enjoying the Knorr pasta sides lately and have found some Mountain House meals good too. There were some Lasagna Mountain House meals in a hiker box that I was able to try and they were delicious. The Mountain House Pasta Primavera is my favorite too. I will also carry out oatmeal, snacks such as fig bars, pistachios, candy, rice crispy treat bars, Belvita crackers, and whatever else seems appealing at the time I'm in the grocery store. Usually the day I leave town I will pack out some actual food like pizza or a chicken biscuit to have for dinner that day.

Trail Angels have been one of the unexpected pleasures on the AT. I have received some amazing trail magic from Trail Angel's along this journey and I'm so grateful. Yesterday I came into a shelter to have a break for lunch and there was a note next to a pack of Mountain Dew that said trail magic help yourself. Those kinds of things are a nice surprise for hikers, especially when we're out here day after day doing long miles and days. Fresh Ground is probably the most well-known and predominate Trail Angel out here since he has his café set up for the hikers and does so much work to follow us and feed us every few hundred miles. That's where I'm headed this morning actually. I've heard he is set up further up the trail.

Someone asked me the other day how many miles or hours on average do I hike in a day. That varies as well. Typically, I will wake up and try to be out of camp by eight-ish and hike to a shelter that's usually 16 to 25 miles away. Sometimes that mileage is shorter and sometimes longer. It also heavily depends on the terrain and difficulty of the day's hike. I can figure all that out by looking at the Guthook app to determine what is ahead of me and the resources. I would say on average I hike about 10 hours a day. Some days are low mileage Nero days as I get to a town to resupply and shower. Usually, I'll resupply about once a week and that will include showering and laundry most of the time, sometimes not.

Day 52.

Well, yesterday was one of my most favorite days on the trail so far and I will be happy to explain why. The day started with a nice 8-mile hike, the last 4 miles of which I hiked with Zoom Zoom, all the way to Fresh Ground's set up. We had such a good conversation that the 4 miles flew by so quickly and I love it when that happens. We made it to Fresh Ground and were greeted by a whole bunch of other hikers and filled our bellies with some good food.

The next 8 miles of the trail are some of the more difficult parts of the trail up to this point. It leads to a peak with beautiful views, called "Dragon's Tooth." Fresh Ground offered to pick me up at a parking lot at the end of the 8-mile stretch so that I can bring my pack but not have to carry as much stuff as I usually would and I would be able to spend the night with the rest of the hikers at his camp. I graciously accepted the offer and two other hikers joined me, "Dine and Dash" and "Squirrel." We set off with a lighter load and had such a great day hiking in beautifully warm and dry weather for a change. We took some pictures and just had a blast exploring the area. Then Fresh Ground picked us up and we went and bought three cartons of ice cream to take back to the camp. Then Fresh Ground made us all a big dinner and delicious ice cream for dessert. After that, we built a fire and all hung around chatting until hiker midnight which is around 9 o'clock, at which point most people head to bed. It was fun joking around, telling stories, and visiting with the other hikers, some who I had not seen in a while. There was 12 of us altogether, 13 including Fresh Ground so it was a nice group and a lot of fun. The perfect ending to a wonderful and beautiful hiking day!

Today I am hiking to another iconic place on the trail called McAfee Knob. It's the most photographed spot on the Appalachian Trail and I'm looking forward to getting some pictures once I get up there. I have about 2 1/2 more miles to go. Not an easy hike since it's at the peak of the mountain but I've been told the view is absolutely worth it.

It's been nice lately, hiking on my own and not having a schedule. I've been hiking consistent miles but it is cool not to have an alarm to get me moving. Funny thing is I wake up at the same time anyway so it works out. The weather is starting to finally warm up and I'm so happy about that! I'm hiking in shorts right now and it feels wonderful.

Day 53.

Yesterday the views on McAfee Knob were absolutely breathtaking. I took a whole bunch of pictures to capture the moment. I didn't think the day could get more appealing being on top of McAfee knob but then hiked over the Tinker Cliffs which were just as gorgeous. You walk along the Cliffs for about a quarter of a mile and it is just amazing. This is definitely part of the trip I would not mind doing again, although a strenuous hike, it was worth it. Just absolutely amazingly beautiful, I'm in hiking heaven. I ended the day coming into a shelter where there were a bunch of section hikers along with some thru-hikers that I knew. The ladies shared their peanut M&M's with us. It was fun to chat with them as I ate dinner and before I went to bed. I had a good night sleep and hit the trails before 8 o'clock this morning.

This morning I'm having a particularly good start. There's a little chill in the air but not too bad. I'm just wearing my rain jacket and a long sleeve shirt and I feel comfortable. I've just finished enjoying a beautiful sunrise as I walk along this gorgeous trail, sipping on my ice instant coffee, instant coffee that is iced from cold stream water of course. I'm also enjoying a wild cherry lollipop as I am listening to an Enya album. Doesn't get a whole lot better than this! Clear beautiful skies, amazing trails, and peace of mind. I know all mornings can't be as great as this but I will enjoy each one I am given to the fullest.

Today will be an easy hiking day since I will hike into Daleville, have lunch, charge my devices, and then go back onto the trail for another 5 miles to a shelter. Tomorrow my in-laws and Matt will pick me up off the trail and spend the weekend with me. I can't wait to see them. Today is Matt's birthday! Since it's his birthday, I am particularly happy to see him tomorrow. He will be 28 years old, I married young lol.

Chapter 11: "Everything falls into place at the right time"

Trail lesson: *One of the reasons I had to come out to the trail is to shake out my raincoat. Let me explain. The raincoat is a representation of my life and the areas of my life. The water droplets on top of the raincoat are a representation of things that I was giving time and attention to that were draining me of energy or were not useful to me. I had to shake out my raincoat in order to focus less on society, focus less on my ego, focus less on social media, focus less on drama that is no good for me, focus less on negative people, focus less on food and activities that were not healthy for me, and focus more on things that were important or are important in my life. Important things like my family, my values, spirituality, healthy foods and habits, a healthier frame of mind, better sleep, more positivity, and the things that benefit me in my life as I move away from things that don't. I came out here to "shake out my raincoat," I discovered.*

Day 57. April 22, 2018

I had a great visit with my in-laws and Matt over the weekend. They came and picked me up from the trail and took me to Daleville VA to spend the weekend. We went out to dinner, and then the next day hiked McAfee knob, the second time for me but I didn't mind. It was a fun hike and we took some good pictures. For dinner that night we went to an iconic restaurant that a lot of the hikers talked about called "The Homeplace." It was as good as everyone suggested with a lot of country food served in big dishes until we were full. I recommend going there to anyone that's in the area. It's only 1 mile from the parking lot of McAfee knob.

One thing that we all had to do while they were visiting was to find a new pack. I hated to have to buy a new pack but the one I had been using has a waist strap that is way too big. I adjusted it to go as tight as it could and it was still sliding down my hips, so I needed to get a different one. It wasn't a problem while I had my winter gear on but with less clothing on, it wouldn't get tight enough. That and I have lost a little more weight out here. We found an outfitter that sold the pack that was just my size with an adjustable strap and I purchased it. That happened just in time and I'm so glad that I was able to find a pack that fits better now.

I didn't want to see my family leave. We all procrastinated getting ready and checking out of the hotel and going for breakfast. I wanted to spend every minute with them as I could. There were thunderstorms coming that evening so I knew I needed to get on the trail and get to a shelter before it started to get worse. After saying goodbye and getting a few pictures, I set off back on the trail. I made it to the shelter right before the thunderstorms started, just in time. This was a cool shelter because it had two levels so set up on the second level and enjoyed watching a beautiful thunderstorm. I was happy not to be hiking in it! My family told me later that they had to drive through a big hailstorm on the way home. I'm glad they made it home safely.

Day 59.

Today was one of the rougher days out here on the trail. It was windy and overcast but the main issue was getting lost today. I ended up going the wrong way on a junction and that led me to a trail that I thought was the AT but it wasn't. Having to go on further down that trail I soon became lost for about an hour and a half. I was quite worried for a little while and ended up slipping in some mud and falling on my left wrist. Luckily, I'm okay but it is really sore. When I slipped in the mud everything was covered in mud, including me and my pack which of course became wet. I was so happy when I did finally find the trail. So that was kind of a challenging instance out here.

When I arrived at the place where I wanted to camp it was right next to a beautiful creek. I used my Garmin to have a nice conversation with Matt which helped me feel better after a rather rough day. He was having a rough day too so we consoled each other. Most days are really nice and beautiful out here but every once in a while, things do go wrong. Just like in life though, it will pass and things always get better. This was one of the most beautiful locations I've camped next to and I was lucky to find such a relaxing spot. After a good night's rest and some food, I'm sure tomorrow will be a better day.

Day 60.

Last night was so peaceful listening to the running water all night. When I woke up this morning I just laid there for a while and dozed back to sleep off and on. I took my time getting ready this morning and making breakfast while enjoying the sunshine and the creek. I just started my hiking day and it's 9:45 am. I have 22 miles to go and I know it's going to be a better day than yesterday. It was nice taking my time this morning and enjoying nature. Plus, it's quite chilly this morning so it's nice to stay cozy and warm in my sleeping bag for a little bit longer than usual.

As far as my spirits and mental attitude, things are going well. It's a beautiful day and I am fortunate to be able to wake up in it each day and enjoy nature. I sure am getting my dose of nature these days. Hiking so many miles and hours a day in and day out leaves you alone with your thoughts quite often. I've taken the opportunity to work through some issues that I've had and let go of, including a lot of negativity. In general, I feel much more uplifted and happier each day. I've come to realize that a lot of people do negative things to others because that is what they currently feel like in that point of their life, or are reacting to situations to the best of their ability at that time. I'm going to make more of an effort in the future to try not to take things so personally. There have been many times where I've reacted negatively toward someone and that was because of how I was feeling at that time. I think I need to understand that about other people and realize some people are having a bad day, bad week, or bad month and I need to not take their negative actions so personal. So that is my goal for the future.

I want to mention that the new pack that I purchased last weekend is working out perfectly. It's very important to have gear that fits just right. I do a lot of slow jogging with my pack on so having something that is snug reduces the bouncing. This pack is great; I wish I had purchased this one at the beginning of this journey. Anyone that's getting into backpacking, be sure you get a pack that fits just right or at least has an adjustable waist strap that will still be effective if you lose some weight.

Day 62, Waynesboro Virginia.

Today was an exceptionally smooth day on the trail. A few days leading up to today I have been fretting over how I am going to make arrangements to get off the trail and get to the airport for my race in California that will be this coming Sunday. I had to figure out where I would be in the Shenandoah State Park, how to get out of the park, how to get back to Waynesboro, and then get to the airport somehow on Friday for my flight at 2 o'clock. Then I was concerned about how I would get back since my flight comes in late, on Tuesday at 11:10 pm, and then how I would get back to the part of the trail that I left, which would be far up the trail. After worrying about it so much I decided to say a prayer and give it to the angels to help me out. Every time I started to worry about it I would just pray to my guides and angels that they would take care of the arrangements and things would happen smoothly. Well, it did and I am utterly amazed at how things worked out.

Let me explain. It started this morning after hiking 5 miles to the exit of the trail near the visitor center near Waynesboro. As soon as I walked out of the woods a gentleman was sitting in a car about to leave, heading in the direction of Waynesboro and asked if I needed a ride. He was not a creepy looking guy, had bike gear and a bike, and said he had just finished biking and was on his way home. I graciously accepted his offer for the ride and he took me to downtown Waynesboro to the CVS since I needed to get a few things. Then while in CVS while I was checking out I asked the cashier what direction the laundry mat was since I needed to do a load of laundry desperately. As she was explaining the directions to me a sweet lady behind me offered to give me a ride. I graciously accepted, and she showed me where I could go to the YMCA for a free shower, and to the grocery store where I could resupply. That was nice of her to take the time to do.

Then I took my time doing my laundry and cleaning up some gear before I walked down the street to a place called Ming. It's a well-known Chinese restaurant along the trail that is known to have the best buffet. It wasn't bad and I enjoyed a great big lunch. Then I walked a short distance to the grocery store and resupplied on my food before setting out to walk to the YMCA which was only about 3/4 of a mile down the road. As I'm walking down the road, with my pack on, of course, a gentleman in a bright yellow pick-up truck that said "shuttle service" pulled over and asked if I needed a ride. I said I didn't have far to go and that I was going to the YMCA for a shower. He said it's not a problem and offered again to give me a ride which I graciously accepted again, such nice people around here! Keep in mind this was a very kind, slightly older gentleman who was not scary at all. I'm not making a habit of just jumping in cars with just anybody, so don't worry.

Before I left to go into the YMCA he offered to come back later and take me 10 miles back to the trail. It was very nice of him and I accepted his offer. Two hours later after a hot shower and putting a few highlights in my hair from the kit I purchased at CVS, I felt like a new woman. I had become so grimy from being on the trail for the past week that a hot shower felt amazing. Plus, the lady at the YMCA was so sweet and nice; she gave me soap and a towel. My new shuttle friend picked me up and drove me to the outfitter so I could get a new fuel canister before taking me back to the trail. Very kind of him, and all for free no less. While we were driving I asked him if he would be interested in picking me up along the trail next Thursday, taking me to the airport the next day, and possibly picking me up and taking me back to the trail. He said he would be happy to and we made all the arrangements for days and times for that to happen. He is going to pick me up on Thursday before my flight, 80 miles up the trail in the Shenandoah State Park which is an hour and a half drive for him, drive me back to Waynesboro, the next day take me to the airport, the following Tuesday pick me up from the airport very late and take me to Waynesboro, and then pick me up the next morning and drive me all the way back to where I left off the trail 80 miles away! All for $100. This is a huge amount of time and driving around and he is more than generous to do this for me. I am in awe of how smoothly all of this worked out. I have my angels to thank for placing me in the right place at the right time to meet the right people and asked the right questions for everything to work out so well. Part of that later included a very nice trail runner named Bill who very kindly gave me a ride in all that mix as well. He went out of his way after he worked all day to help me out and was very helpful. I had met him running along the AT trails, we had a nice conversation and I mentioned a dilemma I had with transportation and he was able to help.

I am back onto the trail from Waynesboro and I am just so relieved that I have a solid plan for how everything is going to work out. Now I get to slow down my hiking pace since I want to preserve my legs for the race and enjoy this beautiful state park. I have 80 miles to hike which I could do in four days but now have 5 1/2 to do them. Nice and easy, relaxing and fun, which is what it's all about. I'm walking along and feeling very grateful and happy, and of course, feel a huge relief.

While I was at the grocery store resupplying I ended up buying something that I did not really need but wanted, that's something I have not done in a long time. I bought a Burt's Bees shimmer lip-gloss. I have not worn any make-up whatsoever, except my Chapstick since February. The thought of putting on make-up, especially eye shadow not only seems foreign to me now but really weird. I put my new lip shimmer on while I was at the YMCA after my shower and felt so beautiful. I'm really enjoying the no make-up feel. The lady that gave me a ride to the Laundromat said that I remind her of her daughter who section-hikes since we seem to be the same age. I then told her I was 36 and she was surprised and said that her daughter was 30. She said I look very young and that made me feel nice. So, there may be something to this no make-up business. Less toxic chemicals on my face may be a good thing in the long run. It sure does make life a lot easier not having to worry about make-up, not that I really did anyway, but it's nice not having to deal with that at all. I am more than satisfied with my lip shimmer and that's it.

Speaking of cosmetic stuff, I would like to mention something else. A few days ago, a relative made the comment, "You don't look skinny." I guess some people have in their mind that thru-hikers will eventually look emaciated, and some of them do. I was not really offended by this statement but it did get me thinking. No, I am not skinny according to some people's standards, but I am very strong. I can accomplish tough hiking mileage in the 20s consistently in these hard-elevated mountains day after day. My legs are strong, my core is strong, and my mind is more at peace than it has ever been. I would not trade any of that to look "skinny" ever. I have been eating well on this trail and plan to continue that. It's been a big change to view food differently now. I look at food as a way of sustaining energy. Sometimes I eat as much as I can so that I can hike as far as I can on difficult terrain. In the past couple of months, I have not once denied anything I wanted or that my body has craved as far as food. It's just a new way of looking at food out here. I know I will have to go back to moderation as soon as I get home but right now it is working for me out here. I feel strong, healthy, and have been taking the vitamins that I carry along with me as well. I am the strong, beautiful me that no one can dictate how I should look.

Chapter 12: Shenandoah National Park and the Badwater Salton Sea 81 mile Ultra-Marathon

Trail lesson: *BLS before ALS. What I mean is referring to something that we use in the medical field. Basic life support before advanced life support. I think this can be applied to all aspects of life. I believe that you have to have a solid foundation in the basics before you can advance into more complicated situation and skills. For example, be a basic good human being and then from there advance on to doing more good deeds for humanity and to help others but you have to start being good and true to yourself before you can help others. Basics first.*

Day 63. May 5, 2018

Today is my second day in Shenandoah Park and so far I'm really enjoying it. Yesterday I had a late start getting on the trail since I had to go into Waynesboro to resupply, shower, and do laundry. I did get 5 miles into the trail and found a nice little secluded spot on top of a mountain with a beautiful view where I camped for the night. I'm impressed with how well marked this park is. It would be impossible to get lost in here, which is a good thing for me, especially recently. My favorite thing about this park so far is all the friendly people. I have passed a lot of section hikers and it's nice to see more people throughout the day. Don't get me wrong, I enjoy the seclusion of the woods but after going all day or most of the day without seeing a single person it is nice when you actually do see people.

I'm planning on hiking as far as I can go since the next couple of days it's supposed to be rainy. Going to enjoy the dry, partly cloudy weather while I can. This will probably be my last big mileage day before the race. I'll start tapering from tomorrow on. (Ended up doing 29 miles that day).

Some information about the race I'm doing next Sunday. It is the Salton Sea 81-mile Badwater race in southern California, in Borrego Springs. It's a teamed race that I will be running the entire race with my friend, also an ultra-runner. He and I had signed up for this race before I committed to doing the Appalachian Trail. I have been looking forward to this race and a five-day break from the trail. I will be flying out to California on Friday and flying back to Waynesboro VA the following Tuesday. Thanks to an amazing Trail Angel, the shuttle service guy, all my arrangements to be picked up from the trail and returned are made. Now I'm just enjoying the trail and looking forward to the race.

Day 67.

Today I am flying to San Diego California to meet up with my friend and some of his family who will crew for us during the race. This morning started out really well and I would like to explain why. I slept well last night at a public park in Waynesboro that allow thru-hikers to camp there for free. Two other hikers joined me last night so I felt safe. This morning I woke up early and went to the YMCA where I was going to meet the shuttle friend to take me to the airport. I arrived very early and was just hanging around and I was thinking about going to get breakfast. A sweet little old German lady saw me and offered to take me to her house and make me breakfast. I accepted her offer graciously and off we went. She was so sweet and kind. She made me a delicious breakfast with the best German coffee I've ever had. She talked to me all morning about how she migrated to the United States from Germany with her husband and 5 sons. How she, even in her 80's, manages multiple rental properties. She is one busy woman and really loves her life to the fullest. It was a pleasure to meet her and spend the whole morning listening to what she had to say about life. After my belly was full and the conversation was had, she took me back to the YMCA and I met up with my ride to the airport.

I am currently in the Chicago airport right now and have been enjoying walking around for the past two hours throughout this huge airport since I have time before my next flight. It's been so much fun to sip on some coffee and people watch. When did the world to get so busy? Guess I've been slowing my life down in the woods for too long because everybody seems to be in such a big hurry, more than I remember. Or maybe it's just this busy city, I'm not sure. Either way, it's quite entertaining since the only people I see these days are smelly hikers. Speaking of smells, one of the first things I have noticed walking around the airport is all the perfume, cologne, and body spray, a lot more than I ever noticed. I can't say that it's bad but can be overpowering at times when you are in close quarters with people.

I'm getting really excited about the race, especially as it is getting closer to the start time! I have not run an ultra-marathon in many months so this will be a lot of fun. Between the new location to race in and the break from the trail, I have been enjoying this side adventure so much. It's been nice to be in the real world after being in the woods for the last 67 days.

Day 70.

The race went great and was a fun challenge! My friend and I had to run the whole race together and were not allowed to separate at any point except to use the bathroom. It started out just fine and then became hot! I wasn't used to the heat at all since I've been up in the cold mountains so long, so I struggled a good bit. My friend, being from Texas had no problem at all with the heat. Then the wind started at about 20 miles into the race, which helped with the heat but slowed us to a walking march for a while. We had a great crew consisting of my friend's sister, her husband, and my friend's son. There were not any aid stations so they would drive 3 miles ahead and meet us along the way, for the whole race! As we grew closer to the end it became every two miles. They were so helpful getting us everything we needed, which was a lot of ice for me in the beginning since I would pour a cup of ice down my sports bra each time I saw them to try and keep from overheating.

At the halfway point, we had an 8.8-mile mountain climb. This was my favorite part of the race by far! My climbing legs were strong and I was thrilled to get out of the intense heat. Climbing up the mountain and over the rocks made me feel like I was back on the Appalachian Trail, so I felt comfortable and strong.

Then it started getting dark and cold. I was used to the cold but my friend struggled with it. We had layers to wear so we were ok. Throughout the night we would talk, walk, run, and meet up with our crew. Around 1 am I hit a wall and became so sleepy. As soon as I made it to the crew I had to lay down for a 5-minute rest and that was just enough time to reset my energy and we kept going. Around 3 am Pete Kostelnick and his teammate Sandra passed us. I was able to meet him in person before the race and was thrilled to get a picture with him. Such a cool guy that I admire very much as an amazing ultra-runner.

We finished the race strong and were happy to be done. I think our time ended up being around 23.5 hours. The sun was starting to come up as we crossed the finish line. By the end of the race, we were all exhausted. There were 40 teams that started the race, 25 teams finish the race, and we were 12th to finish. We were quite content with that and had a great time. We received our belt buckles, picture, and then headed to a cabin they had rented to get some rest. It was a fun race that I would do again one day after a bit more heat training. We were also spoiled with the best crew out there since they worked so hard to keep us warm, fed, hydrated, and going the right way throughout the whole race.

Day 74.

Today is the second day I'm back on the trail since my trip to California for the Badwater race. Yesterday took some adjusting to get back to trail life but today I feel settled in again. Woke up to beautiful weather and have been having a nice hike today so far.

My flight back home went smoothly. Trail Angel shuttle guy came through and picked me up at the airport very late on Tuesday night and had my pack ready for me. I arrived back on the trail the next day to continued my journey.

Day 75.

Today was off to a good start. I set my alarm for 6 am and started earlier than I usually do because I wanted to cover 24.1 miles today, and hopefully before dark if possible. Last night when I made it to the shelter area for the night some other hikers told me that there was a bear in the area an hour before I arrived and it had chased off four other hikers. The bear was nowhere in sight and I was done for the day so I went ahead and set up my tent. Every little noise last night made me wake up and wonder if it was the bear hanging around being curious, but I never did see him. I'm walking along the trail today and it is a beautiful day, slightly overcast, nicely cooler temperature than yesterday. I am amazed at all the wildlife that is popping up now. There are so many chipmunks that run around and play and it's so much fun to watch them and the squirrels, quite entertaining. Today will be my last day in the Shenandoah National Park. It has been a fun park to hike through with all the added hikers and easy trails.

Chapter 13: The "Kissing German"

Trail lesson: *It is all about the journey, so when you get the chance to listen to an elder talk about their past, sit down and listen. The little old German woman I met on the trail in Waynesboro VA taught me a lot with what she said and didn't say. I saw how kindness is lived through the life of this old woman. Enjoy the journey, slow down, and most of all... listen.*

Day 77. May 12, 2018

Well, today something happened that I never would have expected. I decided to stay at a hostel called The Bear's Den Hostel, run by the Appalachian Trail Conservancy. This has been my favorite hostel so far because it was very clean and they gave us pizza and ice cream which was included in our stay. Since it had been raining most of the day yesterday and today I decided to stay indoors to get my gear dry and myself cleaned up.

When I arrived, there was only one other person there who was a young German guy. After taking a shower and cleaning up my gear, he and I had into a nice conversation. He is from Germany and has to go back in two weeks and came out here to hike the trail for a month. He was taking it easy at the hostel because of some bad blisters. Then we talked about various challenges they have along the trail such as the half-gallon of ice cream eating challenge, the Four State in 2 Days challenge, and so on. Then he told me that there was the Kiss A Stranger challenge. I gave him a side look and said I haven't heard of that challenge and he said that it WAS one of the challenges and asked me if he could kiss me. I wasn't too sure about this challenge or him but finally said okay. He immediately jumped up off his bunk bed, walked over to me, and planted a big kiss on my lips. Big surprise! I then pulled away and started to laugh. I went back to cleaning my gear and that was the end of that. Later on, I had to ask him if that was just a line he used to get to kiss me or if that really was a challenge. He smiled and said it really was a challenge that he wanted to complete and he was worried that I would get mad if he asked. I found it funny and we became friends from then on. It was the most action that I've had in a long time, so it did make me giggle about the whole thing. Then he asked me how old I was and couldn't believe that I was 36. He told me I looked much younger and then I asked him how old he was and was quite surprised when he said 19. I laughed again and told him that I met my husband when he was 19 years old. After the kissing incident, I immediately texted Matt about it, of course, and he thought it was funny too. I ended up having the best Nero day hanging out with my new German friend, eating pizza and ice cream, watching movies, and cleaning my gear. Just as a Nero day should be.

My husband and I are comfortable sharing experiences we have with each other since we opened our marriage to a polyamorous dynamic about 4 years ago. We support each other having new friendships and experiences with other people when opportunities present themselves. That is why I dated my ex-boyfriend for 2 and 1/2 years as our first polyamorous relationship. My husband has dated also when he finds the right person and situation. Bottom line is that neither of us keeps any secrets from each other and always are supportive of one another. We understand that many people do not understand or agree with this type of relationship and that's fine. We do what makes us happy and don't hurt anyone else.

Day 78.

I am almost finished with Virginia! It has been my favorite state to hike through by far. Virginia is absolutely beautiful with the most amazing trails. I hope the rest of the Appalachian Trail is this much fun. Tomorrow I will be at Harpers Ferry, which is the unofficial halfway point. I could go past that town today but I want to stop at the Appalachian Trail Conservancy and have my registration recorded and picture was taken in their office so this hike can be official, a little more official at least. Tonight, I will be hiking to a campsite near a town so that tomorrow morning I can hike into Harpers Ferry. I'm very happy that the weather has started to improve. It's not so deathly cold anymore and the snow is officially gone. It's nice to start hiking in the morning without everything being numb from coldness. Now I think the biggest challenge will be rain.

Day 79.

Last night I stealth camped alone at a campsite on the ridgeline of a mountain, peacefully. Woke up this morning and hiked four miles to the Appalachian Trail Conservancy headquarters. It was nice to arrive and get my picture taken and have it added to this year's thru-hiker 2018 book. I felt so official. I even received a 2018 Appalachian Trail marker which I attached to my pack. They give those to official thru-hikers and it's a different color each year. I am the 126th hiker to make it to Harpers Ferry. It was fun to look through the book and see all of my other hiker friends and when they had come through as well. Surprisingly, we are all closer than I had imagined to each other.

While I was at the Conservancy one of the volunteer ladies started a conversation with me. She wanted to know how my hike was going and asked a few questions about it. She said that female hikers have a special place in her heart and that she was proud of me for doing this. She urged me to be in contact with them when I finished the hike and to write an article about my experience for the Conservancy. I told her that I would be happy to. After a nice big lunch and re-charging of my devices, I had a beautiful walk along a big river before going back into the hills. I am in Maryland now and Virginia is behind me. Virginia was so much fun that I hate to have that section end. The last 200 miles of the trail has been my favorite so far. My favorite because of the beauty, the well-marked trails, the nice people, and clean hostel I stayed at before going back into the hills.

I just saw my first snake on the trail. I knew I would see one but I didn't know when. This little guy was a beautiful and bright green color. I also saw a whole bunch of turtles lying on the rocks and tree branches that were in the river as they baked in the warm sun. Summer has officially arrived, it's about time.

Day 80.

Last night I slept so well at a campsite in Maryland. It was a designated campsite that had bathroom facilities which included a shower with warm water. I saw this place on the map so I had brought some shampoo and soap that I found in a hiker box at the Conservancy earlier in the day. It sure was nice to get a hot shower before going to bed.

Day 81.

This morning I woke up to a woodpecker going to town on a tree right above me. It was all right because I wanted to get up early anyway and get an early start on the trail. Some thunderstorms were coming my way and I wanted to get as many miles done as I could. Luckily, I made it 8 miles in before it started to pour down rain. This is my first torrential downpour with thunder and lightning since all the snow earlier in the trail. I passed a big group of camp kids that did not seem too thrilled to be out in the heavy rain.

Yesterday I crossed the Maryland/Pennsylvania border. Maryland only lasted two days and had beautiful trails.

Chapter 14: Baby Goats and the Hospice Woman

Trail lesson: *Passion. I believe everyone is passionate about something. If you don't know what it is then you should figure it out. While out on the trail I asked a bunch of people what they were passionate about. The people that were doing something towards their passion for what they are passionate about tended to be happier people in general. I think there's something to it. I am passionate about running and spirituality just to name two. The more I have embraced the two of them and enjoyed it the happier I have become.*

Day 87. May 25, 2018

I would like to start out by saying I had a wonderful visit with Matt over the weekend. He drove eight hours, all the way from Charlotte to pick me up in Pennsylvania and take me the two hours to his grandparents' house, where we visited with them and his dad over the weekend. He even took off Monday so we could spend the day together. It was a perfect visit. Yesterday he took me back to the trail, exactly where I left off. I am more determined now than ever to conquer this trail and get home. It was difficult to leave him yesterday and I miss him terribly. Of course, I'm having the experience of my life out here but I do miss the comfort and routine of home, especially since I haven't even been there for 87 days.

Just for the record, I did not consider going home or quitting the trail. I've come way too far and had so much personal growth to turn back now. I do know it's about to get hard though. New Hampshire and Maine are quite difficult hiking but I will be as ready as I can be for it.

Yesterday I was caught in a torrential rain and hail storm. I was pelleted by these balls of hail that were coming down so fiercely! I was in an open field for most of the storm so that was fun as the thunder and lightning lit up the sky around me, said sarcastically. I make it to the shelter, which was full, so I set up my tent nearby. I would have tented anyway but it was nice to have lots of other people around. It rained early this morning and is supposed to rain all day today. I'm ready for it in my rain jacket and rain skirt. I'm hoping to knock out 22.3 miles today. I don't mind the rain for the most part and find it mystical and soothing. This morning I started my hike early and there is a strange ambiance in the air. It's drizzling, dark, and very mysterious looking out into the forest. I almost feel like I'm in a movie as

I'm walking along this path, deeper into the woods on this rainy morning.

Day 88.

I know I like to tell you all the good parts of the trail for the most part but I need to complain for a second. One thing that I do not like out here is having to put on wet socks that are cold and soaked in the morning after it has been raining for literally over 24 hours. I guess I would take that over frozen socks but it's still no fun either way. I have three pairs of socks that I switch out and currently, two of them are soaking wet and one of them is dry. There's no point in putting on the dry pair because my shoes are soaked, insoles are soaked, and it is still raining of course. Luckily, after I have them on for a few minutes and start moving around it's not so cold and miserable. When I finish the trail, I will not miss the wet sock mornings.

On a brighter note, I slept well last night in my tent and it seems like I woke up in a rain forest. It's foggy and drizzling steadily. It is quite beautiful and majestic to wake up to and walk in. It's not pouring down rain so that's a plus. My tent is soaked and so is my pack which makes everything a lot heavier. Looking forward to some dry days that will eventually have to come, right?!? I looked at the weather and it's supposed to rain for the next four days so eventually, I will have dry socks but probably not anytime soon.

This morning I made a Facebook post with the main message being how much I appreciate my family and friends. I never realized how much I enjoyed seeing the people that I love and care about consistently and so easily. There is a lot of alone time out here which I do enjoy but it is so nice to meet up with somebody for coffee and conversation or to visit with someone that I care about. I miss that and look forward to seeing all my friends and family when I get finished with the trail. It makes me more determined than ever to finish this adventure and get on home. I miss Matt and Millie (my dog) the most and look forward to being home.

I'm hiking along today; the rain has stopped and the birds are starting to come out and play. I was listening to a Ted Talk that I have downloaded that discussed the concept of various perspectives on life. The girl giving the speech went around and asked random people what they wanted to do before they died. People had various answers; most of them didn't know what to say and were caught off guard by the question. Then one gentleman said that he would hike the Appalachian Trail. That struck a chord with me when I heard him say that because that's exactly what I'm doing. The more I thought about it the more I realized I don't have a bucket list or anything I have left undone before I die. Once I have the idea, I go and do it. I'm not saying that everyone needs to go after every idea they have with full force but I think people should consider doing it more often if they can. I feel bad for the people that can't answer that question and even worse for the people that have so many things in their life that they want to do but haven't. For some reason, it made my eyes swell up with tears when I heard that guys answer. I'm not sure exactly why I feel so emotional, but I do know that I am overwhelmingly grateful to be doing what I am doing right now. There are difficult days and moments for sure but isn't that how life goes? With the good and the bad come opportunities, opportunity to grow and learn. I'm embracing those opportunities and hope that from the bottom of my heart more people will too. The TedTalk was called, "How to skip the small talk and connect with anyone."

Day 91.

Yesterday was a tough day with all the rain. It was 45° in Pennsylvania and I hiked through a day where it rained most of the time. I was soaked down to my underwear and my hands stayed cold almost all day. Things did turn around though. I ended up getting to a town called Port Clinton Pennsylvania where they had a restaurant. I went in there and had a nice dinner and met another hiker there. He had reserved a hotel room and felt bad that I was going to go back out into the rain to pitch my tent and camp. He offered for me to stay in his room and I slept on the floor on my sleeping pad. It allowed me to get a shower and dry out my stuff which was incredibly nice of him. I was very thankful. He was also a complete gentleman and treated me with a lot of respect, meaning he did not hit on me or anything. I had a good night's rest and started the next day fresh, dry, and clean.

This morning was off to an amazing start. I told the hiker who was so kind as to let me stay in his hotel room that I would like to treat him to breakfast for his kindness and letting me stay in a dry room. We were walking down the road to go get breakfast when I was telling him about Fresh Ground. I was telling him how he is one of the main trail angels out here and how he feeds all the hikers when he can along the trail. I'm not kidding when just a couple minutes later we walked past a pavilion and there he was! Fresh Ground was there cooking breakfast for all the hikers that had accumulated there. We stopped and had an amazing banana pancake breakfast as usual. It was so nice to see him and even better that it was a surprise. I thought he was a couple of days ahead of me and I was pushing it to try to catch up to him so this was a very nice surprise.

After chatting all morning and eating pancakes, I was in a great mood and felt ready to tackle a hard day of hiking. The kindness of that hiker to let me get a nice dry and deep sleep along with the generosity from Fresh Ground, I felt fantastic! Out here, I realize that I am a strong individual but I am even stronger with the help and kindness of others. I'm fortunate to have so many kind people in my life, whether they are people I know or strangers.

I'm also realizing out here that there are tough times just like in life, but they don't last. You're going to have rainy cold days but it will soon be followed by a warm sunny day. Just keep going forward and it gets better.

Day 92.

I almost stepped on a Rattle Snake! There was a big log lying across the trail and as I took a big step over it, I started hearing the rattle sound. Then I noticed this big snake scurry away from my feet. That was a close one! Yikes!

Day 93.

I am pleasantly surprised to find out that the last video I made which had an inspirational message in it, was well received by a lot of people. I made the video during that rainy day when it was quite miserable and cold out, not thinking too much about it. I posted it to Facebook the next day and was pleased by all the positivity I received from it. It sure was not expected but made me feel very happy. It did get me thinking about a lot of important concepts and lessons I come to realize out here that I tend to keep to myself. Maybe I should not be concerned with that as much and share more. We will see what comes of it. I definitely don't want to get too wrapped up in social media out here but it is nice to share some concepts I've learned while hiking with others, especially if it can help people.

Day 94.

The Pennsylvania rocks are quite a challenge! The first 2/3 of Pennsylvania are not too bad and I thought that I can handle this with no problem, and then the boulders appeared. The last couple days, I've had to put my trekking poles away and use both hands to scramble over huge rocks, along with rock fields of smaller loose rocks. It's been quite a challenge and I've fallen a few times. Falling on hard rock is not as fun as slipping and sliding in the mud and snow. I'm finding that rocks tend to leave more bruises. That's okay though, it's making me tougher. A lot of the rocks are sharp and jagged and you have to step on them since it is the only path. This has caused a lot of bruising on the bottom of my feet too, so that's been a challenge.

I was pleasantly surprised when two of my family members, that I used to see at my family reunions, reached out to me not long ago and told me that they lived in Pennsylvania right next to the Appalachian Trail. I didn't even know this. They offered for me to stay with them as I was passing through, which I did. It was perfect timing because the night and day I stayed with them were a lot of thunderstorms and rain. It was so nice to see them and catch up as well as clean and dry out my gear and get a nice shower. These past few rainy days have been a challenge so it was a very welcomed break to stay indoors for a night.

They were very kind and took me out to dinner and also let me play with their three little goats. They are raising three goats that are full of energy and character! They just turned a year old and needed their CDT shots. I offered to give them their shots since I have plenty of experience giving people shots and also animals when I used to train horses. They didn't even notice their shots because they were drinking their bottles, LOL. It was so much fun to play with them!

Also, during my visit with them, we went to see my relatives' mother who is in hospice. She is on her final days of life as her nurses have stopped all her medications since she can no longer swallow or speak and are focused on making her comfortable. It puts things into perspective that life is so precious. It was an honor to see her in her final days, as they don't expect her to be on this earth much longer. I hope she is at peace and not in any pain. She ended up leaving this life the following day with her two children by her side.

Visiting hospice has me thinking today about life, not that I don't think about life every day anyway. It was interesting to see her room, very simple with photographs of her family surrounding her. Of all the things that we accumulate in our lives, what is really important at the end of it? All the memories and photographs of our life? Makes me feel that I need to put a lot less emphasis on things and much more emphasis on experiences. After this trail adventure, I intend to be a lot more focused on creating good memories with the people I care about.

As a side note, I have incredible respect and admiration for hospice nurses. I believe they are some of the kindest and sweetest humans on earth. In my opinion, they not only care for their patient's human bodies, they care for their souls. As an empath, I could feel all the love in that woman's room while I was in there. What an incredible and peaceful place to transition at the end of your life.

A few days ago, as I was hiking the trail, I would not have expected that I would have an incredible experience in a hospice facility. As always, I am grateful for the opportunity to have been there and meet her. It goes to show that you never know what life will present to you, but being flexible and going with the flow, allows opportunities to enter our lives in many different ways. This is just a small example of that. I gave that sweet woman a hug before I left and said a prayer that she passes peacefully, which she did the next day with her family by her.

Day 95.

Out of Pennsylvania and onto New Jersey. Very happy to leave those awful rocks behind me. That is all.

Day 96.

I understand the cycle of life and how it works in nature, but I have to admit that it's not always so easy to see with my own eyes. Maybe with the passing of that elderly woman has my emotions triggered, I'm not sure. I was just now walking along the AT path when I look to my left and noticed a big snake that had caught a frog. The frog was still alive and the snake was busy consuming it. The frog looked so scared and helpless. I could not save him nor think I should disturb the food chain. I will admit that I walked away and had tears in my eyes for that poor frog.

Although the cycle of life is hard to witness sometimes, it is necessary. That will give me something to think about today. Goodness, these lessons keep coming lately!

Chapter 15: It's My Birthday. 600 miles to go!

Trail lesson: *I have realized that it is important in life to try to be more present and enjoy the moment. About halfway through the hike, I found myself stopping to take in the sights, a feeling, an emotion, and experience. I realized in those moments that I am exactly where I need to be at that moment. By doing that, it made me feel good and I began to do it more often. The more often I did it, the more present I became, which made me less worried about the past, less worried about the future, and more appreciative of exactly where I was and what I was doing at that exact moment.*

Day 97. June 8, 2018

I finally saw a bear! Wasn't sure if I was going to see one or not, but it finally happened. Last night around 8:30 pm, the sun had just gone down and I was doing a steady jog, trying to get to the shelter that was a mile away, before running out of sunlight. I was slowly jogging along when all of a sudden, a medium-sized black bear took off running into the woods away from me. We both must've startled each other because I just froze and watched him run away. I had no idea that bears could move so fast! I could feel the ground trembling slightly as he ran away. This was near some blueberry bushes so it's no surprise that he was there enjoying a meal when I came running up and scared him half to death too. I'm excited that I finally was able to see one!

Day 102.

A couple of days ago I was really starting to hit a wall here on the trail. My feet were hurting more than normal at the end of the day and I was just starting to feel down about the monotonous trail schedule every day. Luckily, I had a wonderful visit from a good friend of mine named Richard, who lives in Massachusetts. He drove all the way out here to spend time with me, treat me to a resupply, a hotel room, a long soak in the hot tub, and foot rubs for my sore feet. We even went to REI where I was able to exchange my sleeping pad which had a hole in it and was now useless. This was a good opportunity to get new shoes as well since the last pair had worn down to almost nothing on all the Pennsylvania rocks. The three nights prior to his visit I had been sleeping on the hard ground and not sleeping very well because my sleeping pad would completely deflate each night shortly after I blew it up. It was just the reset that I needed and I'm feeling confident and strong out here again since his visit. The day he arrived was a day that I felt like this trail is never going to end and needed a break. I love being out here on this trail and it's been life-changing for sure, but it is quite lonely and a lot of time to be alone with your thoughts. Which is usually a very good thing but having people to associate with is also something I missed doing.

I am in New York now and just crossed over the Hudson River Bridge. It was really beautiful and I was glad I was able to walk across it during the day.

Later that day… I was hiking along when all the sudden I came to a road crossing and there was a bunch of emergency vehicles. As I approached they told me to be careful on the trail because a hiker had slipped, fallen and broken her ankle. I continued my hike and in about a quarter of a mile into the woods, up some very steep rocks, I came upon the rescue crew with the hiker in a stokes basket and her right ankle wrapped up. I stopped to offer my help, even though there was not a whole lot I could do. She was grateful to talk to me for a little while and then I proceeded on my way. They were going to have a tough time getting her out of the woods considering the terrain. I thought about her for the rest of the day and sent positive energy to her as well as the rescue crew.

Day 104.

Today is absolutely gorgeous outside! There was a beautiful lightning storm last night at 11:30 that I was able to enjoy since I could see it through my tent. It didn't last very long but it was spectacular. This morning I was up early to start my hike and the first part of the trail is walking through these gorgeous rolling hills and fields. There is still condensation from the rain but the sun is shining through and it's so lovely. I've been walking along for a while now and just look around and cannot believe that I'm able to be here right now and see all this beauty. I also passed one of the largest oak trees I have ever seen in my life.

It has been quite strenuous out here and my legs and feet are very sore. Despite the soreness, I have been able to bust out some decent mileage each day. Yesterday I did 25 miles and today I will do 22. As much as I do absolutely love it out here, I am looking forward to summiting Katahdin and relaxing for a few days afterward. Yesterday, due to the difficulty in climbing some of the terrain, it took me 13 1/2 hours to hike the 25 miles. That was with one short lunch break and two other five minute breaks to sit down. I do love the challenge out here but it for sure is a challenge. I seem to be in better spirits lately with looking forward to seeing Matt and my family soon. It will be so nice to spend time with people that I love and care about after this adventure has been completed and I really look forward to that and think about it quite often throughout the day.

Later on that day... I just passed some section hikers and as I looked at them my first thought was, man they're clean. I find it funny because I have been profusely sweating for the last four days in the same T-shirt I'm currently wearing. My plan is to get to the shelter I'll be staying tonight and wash my shirt in the creek. Until then, I think I'm attracting every insect within a 5-mile radius.

Day 107.

Today is my 37th birthday and it is off to an incredible start. I slept really well last night and was up early so I can do a tough climb, which was at the highest peak in Connecticut. The whole way up, I thought about how fortunate my life is and all the great opportunities I've been lucky enough to encounter. I made a video on top of the mountain and posted it on Facebook since I had good cell service. A few things that I was thinking about on my way up on the summit were how I have an amazing life and I've created the life that I want to live. That, and I have a great relationship that Matt and I have worked hard to create, and I'm very proud of it. I have a lot of amazing people in my life that I surround myself with, and I believe that contributes to my happiness and success. I don't have a bucket list of things I want to do because I make things happen as I want to do them.

I'm very much so looking forward to a lot of big future plans. The 200-mile races will be my next big challenge after this trail is completed, I am a little nervous and excited about them. I have a lot of fun plans with people in my life that I look forward to doing after the trail is complete too. I am also considering a career change but that is still in the works and we'll see how that goes.

A little bit about my progress on the trail. I made it through New York and Connecticut and I'm in Massachusetts at the moment. I can tell I'm getting closer to New Hampshire and Maine because the climbs are starting to become more difficult again. I have about 670 miles to go until I am able to Summit Katahdin! Little too soon to be excited about it but I definitely have my eye on that mountain.

Later on that day... I received a text from Fresh Ground saying that he was going to wait for me at a trail crossing when I come by. He wanted to make me a big dinner for my birthday. It was so kind of him to do that, and I hiked eagerly to get to his set-up place. He made me so much food and even charged up my battery pack completely. I had mentioned that I had not showered in quite a while and he decided to help me out with that. He boiled a big pot of water, hung a tarp off the trail in the woods, and gave me a washcloth, towel, and soap. It was so nice to get cleaned up and feel nice and fresh. Fresh Ground really goes out of his way to help **ALL** the hikers out here as much as possible, and he was determined to make my birthday special, which he really did. It was an amazing day and a very strong hiking day.

Day 108.

After such a good day yesterday, today did not go as smoothly. I'm not sure what exactly I ate yesterday but I think it was too much grease for my system and I had an awful stomach ache for most of the day. This was the first time on the trail that I actually felt ill but I used it as an opportunity to hike through it in preparation for the races and not feeling well during them. I was quite dehydrated and had completely lost my appetite. I felt so exhausted all day and every little hill felt like a huge mountain to overcome. I ended up telling Fresh Ground that I think it was something I possibly ate the day before, maybe the eggs, maybe the oils, but something was not agreeing with my stomach. I didn't want other hikers to get sick. He felt so bad and ended up doing a massive bleach cleaning of all his equipment just to make sure there was no contamination. He is really good about keeping everything clean and bleached so I really didn't think it was any germs related to that. I have been super careful out here on the trail with germs. For that reason, I don't sleep in shelters when possible or eat on picnic tables at all. I always sleep in my tent and prepare my food carefully as to not get germs from other hikers.

He ended up setting up his food station at another road crossing today and I told him not to wait for me because I was moving so slowly since I wasn't feeling well. He insisted on waiting for me and as soon as I arrived he had a big bottle of Gatorade waiting for me. He's always so thoughtful. I was able to get some food down and as much Gatorade as I could and felt significantly better. I ended up hiking 20.3 miles today to one of the serenest places I was able to spend the night at so far. It is called the Upper Goose Pond Cabin. It's a big cabin a half a mile off the trail that is run by the Appalachian Trail Conservancy and has a volunteer that stays there to maintain it. It's a big indoor cabin with bunk rooms, a wash station, a privy, and hikers can use the canoes on the lake if we wanted to. It is all free for hikers. I was way too exhausted to go canoeing but I did fall asleep shortly after 8 o'clock and had the best night sleep I've had in a long time, on an actual mattress. It was wonderful! The next morning Pete the volunteer, was up at 6 am and made all of us a blueberry pancake breakfast and coffee. He was so sweet and really takes a lot of pride in maintaining the cabin and trails. We had a nice conversation last night and this morning about all of that. Before I left this morning to start my hike he made a comment that struck me as odd since it was out of the blue. Pete said, "I think you will be famous one day and I will be happy to say that I got to meet you". I smiled and giggled at him and then went on my way to start my hike this morning. Very funny that he would say that and I've been thinking about it this morning. I definitely don't need fame but if I can help people I will feel I am following my life purpose. I think some people can sense that that is my path, like Pete.

My hike is going very well this morning and I'm so happy that my stomach is not feeling nauseated. I'm reflecting on the fact that the bad days just enhance the good days and I need to remember how good it feels when I feel strong. I was able to eat four pancakes without any issues so I was happy about that, and think the carbs will do me good today on this hike. I will be hiking into a town called Dalton, Massachusetts where I will spend the night in my tent in a trail angels' backyard that welcomes hikers to do so. Then I plan to resupply the next day before I head out.

A little bit about the wildlife on the trail. The trail is covered with these beautiful bright orange lizards that are so cute. Especially when it's warm out, they are all over the place and I love seeing them everywhere. I have also seen a lot of snakes on warm days. Only one poisonous one so far but the rest of them are just so beautiful with bright colors. I enjoy seeing them and have always liked snakes, as long as we both respect each other's space which is the case out here, at least so far. One animal that I have not particularly been a fan of in the past are birds. I always found birds rather annoying but being out here and seeing them every day I have a whole new respect for them. First thing in the morning, at the first sign of light, they are out chirping and singing and it is a welcomed sound now. They sing and play all day and as soon as the sun goes down they stop. I have grown to love their singing, especially in the mornings. Of course, I see a lot of deer, turtles and such out here, but the lizards, snakes, and birds are my favorite.

Day 109.

Last night in Dalton Massachusetts, I was very productive. I made it to Lavardi's house and he let me set up my tent in his back yard. This gentleman has been allowing hikers to do this on his property for years. Then as I was walking through town to find the community center for a $3 shower, a Trail Angel stopped me and offered to drive me everywhere I needed to go. I had a much-needed shower, resupplied at Wal-Mart, and bought a burrito for dinner. The angel made it much easier to get around and get everything done, which is so kind. I was back to my tent by 7 pm which gave me enough sunlight to wash all my clothes with a hose and clean my gear. Felt great to have everything clean and resupplied.

This morning I'm having a nice breakfast at a cute coffee shop before I head out of town. I'll do another 20 miles today and have two big climbs to conquer so I'm trying to have a good breakfast to get me through it. I have 600 miles to go after today! Yay, getting closer!

Chapter 16: Into the White Mountains We Go...

Trail lesson: *Sleep. I am sleeping better out here on the trail than I ever have in my whole life… when the intense snoring is not an issue of course. I thought long and hard about why that is the case and finally figured it out. I worry significantly less than I ever have in my adult life at this point. Having less worry and stress allows me to sleep soundly. I fall asleep right away, stay asleep all night, and wake up rejuvenated with an abundance of energy. No melatonin or anything related to sleep aid needed. Better and sounder sleep is a great side effect of less worry and stress in my life.*

Day 113. June 23, 2018

Today's destination for the night after my hike was different and fun. I was able to hike to the top of Bromley Mountain which is the peak of a ski resort. At the top, there is a cabin, which they use for people to warm up in at the top of the mountain during ski season, open for hikers to use during the offseason. It was well insulated and very nice and warm inside, a great way to sleep indoors for a change. I slept right next to a big window and was able to watch the most amazing sunrise as it came up over this huge mountain. Another section hiker showed up before the sun went down that evening and slept in the next room.

I definitely know that we are getting closer to Maine since the hiking terrain has become a little more strenuous lately. My day is filled with a lot more climbing over mountains than it has been. Because of this, I have had an increased appetite. I was so hungry the other day that all I could think about was food, which made it difficult to get into my hike. I ended up making an unexpected trip to a nearby town to buy some food which settled my stomach and my mind. I'll have to keep this in mind as we approach more difficult terrain.

Day 114.

With a few extra calories and some good rest, I ended up having a strong hiking day. Even though there was a lot of climbing involved, I ended up doing 26.7 miles. I have been doing 20-mile days every day for a little over a week now and feel determined to get this trail completed. I'm trying not to overdo it and I am listening to my body and soreness. I pretty much hike all day, usually 10 to 12 hours, and then spend the rest of the time off my feet and resting. I will definitely slow it down in New Hampshire and Maine as it becomes more difficult, but for now, while I can, I will push it.

Day 116.

Today there was a lot of up and down hills to tackle. It has been like that for the last few days so my body is really feeling it. I am feeling worn out, more so than usual. I believe this is due to doing between 20 and 25 miles each day. I have not had a day off in over three weeks and am pushing it because I want to finish the trail in time to rest up before the races start in August. Because I have been pushing it so hard and not sleeping very well each night, I decided to take it easy today. I hiked 15.4 miles to a cabin called Lookout Cabin which is nestled on the top of a mountain. It is a cabin owned by a family that has it on their private property the AT goes through, and they graciously leave it open for hikers to use. This was perfect timing in a perfect location and I took advantage of the opportunity to rest up. It was a hard 15.4 miles yesterday and by the time I made it to the cabin, it was starting to drizzle and get cold. I took the opportunity to eat a good dinner and go to bed early. Since it was enclosed and warm, I slept so well. I slept a solid nine hours without the fear of any animals coming into my tent, thunderstorms, trees falling during the storm, torrential rain, or anybody else around to bother me. There was another hiker there named "Shark Bait", very nice hiker that I had met a couple of weeks ago. The cabin had a loft and I stayed at the top while Shark Bait stayed at the lower level.

After a much-needed nights rest and recovery, I am having a strong hiking day today. Today is Friday and I will be meeting my friend Richard on Sunday. He is going to drive from Massachusetts to meet up with me in New Hampshire. I will still hike that day but not as many miles so that I can have a rest and recovery day and also to re-supply, do laundry, go to dinner with him and get a good night's sleep in a hotel.

I'm excited that I have less than 500 miles to go. As of last night, I had 471 miles exactly left to complete this adventure. I see the end is in sight and I'm excited. Everyone talks about how amazingly beautiful the White Mountains are and I am anxious to tackle them. Well, I know they'll be a good challenge but I'm excited to see their beauty and get to Katahdin.

Day 118.

I had planned to do a 24-mile hike that day. Towards the end of the day, I ran into a trail angel that gave hikers cookies and bagels. After being loaded up on some carbs I felt rejuvenated. I made it to the shelter I intended to stay at and still had some energy and daylight to keep going. I ended up hiking another 3 miles and stealth camped in a beautiful field by myself, peacefully. I can tell that I am becoming a bolder hiker now because at the beginning of this adventure I never would have continued down the trail without a definitive plan of where to stay. Now I feel a lot more confident and know that I can find a safe spot along the way anywhere on the trail, even if it's away from people and completely on my own, which I have done many times now. This is what they refer to as stealth camping. I've done that quite a bit and almost prefer the solitude as opposed to a full shelter of people. It's fun to keep going and see where I end up. Although, this will probably be one of the last times I do this because I'm about to get into the White Mountains of New Hampshire. The White Mountains are notorious for being very difficult and having severe and unpredictable weather. For these reasons, I plan to stay in shelters or huts and be sure that I am there before the sun goes down. I enjoyed the stealth camping while it lasted and I'm happy that I'm not a big chicken out here anymore.

Day 119.

Today's an exciting day because I'm going to meet my friend Richard from Massachusetts who is going to come and visit me. We decided on a meeting place which is at the base of Mount Moosilauke which is the first mountain of the White Mountains. (Turns out I became dehydrated in the high heat and he picked me up closer to my location than planned). As of yesterday morning, I had 44.3 miles to go to make it to that meeting place. This will be the last couple days that I will be pushing the mileage limit and trying to go a little harder. The day before I did 27 miles. Yesterday I did 29.1 miles. Today I have 17.3 miles (did 11.5) to go and meet him around noon. Because of this I hiked last night until 11:30 pm and was up at 4:30 am to knock out the rest of the miles. It takes so long to do these miles because there's a whole lot of climbing to do and that takes so much more time.

Getting up so early this morning was absolutely beautiful to be at the top of the first mountain after the sun had just come up. I stopped for a moment and just gaze over the mountains and soaked in the peace and beauty of it all. The air is so cool and the woods seem to just be slowly and happily waking up. So far, I'm really loving New Hampshire!

I have to mention something that's going to sound a little silly. One of my favorite things to see along these trails is the cute little orange lizards everywhere. I would see a whole bunch of them along the trail up to this point and every time I saw them it would give me a warm fuzzy feeling inside. I think they're adorable and it made me happy to see them. As soon as I walked into New Hampshire yesterday and hiked over 20 miles I did not see one lizard that I love so much. This morning, after admiring the beauty of the mountains, I continued walking and saw a lizard in my path. That has made me extremely happy and a great start to this beautiful day. Wish I could take some of these cute little lizards' home and keep them as pets but I know they belong out here in the woods where they can thrive. That may seem like a silly little thing but it's the silly little things that we appreciate in life that when we focus on them just enhances our joy. Those little guys have brought me a lot of joy.

Day 120.

Today was a big day because it was the first day I entered the White Mountains. I had a wonderful visit with Richard, who drove a long way to visit with me and picked me up from the trail. We accomplished so much after he picked me up. I did my laundry, had a very much-needed shower, a fantastic dinner which was a humongous hamburger and sweet potato fries, a great nights' sleep after soaking my sore muscles in a hot tub, and a foot rub. Of course, the conversation and catching up is always fun with Richard too. We also went to Wal-Mart and purchased some supplies I needed for the next leg of the trip, which included a rain poncho and sweatshirt that I will need for the colder weather in the Whites.

After a lot of thought, I changed up my strategy for the trail from this point forward. I decided to leave my tent and cooking system along with a few other small items with Richard to lessen my pack weight. It turned out that I saved about 7 pounds in my pack by leaving those items with him. That's significant! I have been struggling with good food lately since the hiker food that is available to us and that I cook is not very high in nutritional value. I have noticed it the last week especially with my energy levels being significantly less. I know a lot of that is due to the higher mileage I had been pushing lately but I don't feel the power after eating these foods so I know that is one of the issues. By having less pack weight, I will be packing out more calorie dense and nutritional foods in hopes to maintain my energy levels and calories in the Whites and to finish the trail. From the point where Richard picked me up, I have exactly 402 miles to go. That's very exciting!

After a big breakfast and some coffee, Richard dropped me back off at the trail. I hiked the 6.3 miles to the base of Mount Moosilauke, which is the first large mountain of the Whites. I made it to the top and it was amazing! It was so gorgeous as a huge thunderstorm was coming my direction. I still had to take a minute to just marvel at how amazing it was up there. Then I hiked the rest of the way to the shelter I stayed at. I still was caught in the rain for the last hour but that was fine. When I arrived at the shelter there was no one there so I had the whole place to myself for the night.

It was awesome watching an incredible thunder and lightning storm for most of the night from the view of the shelter. I was very happy to be inside a shelter with all the rain that came down. I will have to be in a shelter or hut for the remainder of this trip since I no longer have a tent. Although, I do plan to cowboy camp at least once before I get to the end. Cowboy camping is when you do not have a tent and you just sleep in the open air in the woods with your sleeping pad and sleeping bag. There are lots of places along the trail that I think I'll be able to do that and I hope to enjoy that one night.

Later that day… I made it to the Kinsman pond shelter after a strenuous day of hiking. I knew the Whites would be difficult but it just took me 11 hours to hike 13 miles. I did take a lunch break and a couple of snack breaks but that didn't last very long. It is so strenuous because I have to put my trekking poles away and use both hands to climb over boulders. Some of the miles are straight up these huge rocks and it takes FOREVER. Talk about being worn out! These White Mountains are majestically gorgeous but they are going to be a challenge to get through for sure. A lot of the rocks are very wet and slippery so I have to take it slow, so I don't fall. We'll see how it goes.

Day 123.

This morning I was up early and started my hike at 5:30 am because there was someone snoring in the shelter that kept me up throughout the night and I was ready to go. It all worked out though because I was able to get a head start on the day's hike and hiked the 3 miles to the Galehead Hut. I stopped in to have some tea and check out the weather report, which is important to keep an eye on up here since the weather can change quickly and become very dangerous. All the huts in the White Mountains have an updated weather report available to hikers to check out at any time. I arrived just as the paying customers were finishing breakfast. I said good morning to the staff and told them I was a thru-hiker coming to have some tea and check out the weather. They were all very nice and said that if I waited 15 or 20 minutes that I could have leftover food from breakfast if I wanted it. No thru-hiker I have ever met would turn down food, and that includes me. After waiting around for a while and chatting with some other hikers, the staff allowed me into their kitchen to have breakfast with them which was so fun. It was fun to chat with them and hear about how they run the operation of the huts. They listen to this radio that is constantly playing where all the huts communicate with each other through an operator. It's old-school and it's really cool to see and hear.

After filling up on a good breakfast, I tackled the steep climb up Gail Head Mountain. Luckily, it's not a long climb but very steep. I arrived at the top and it was a little too cloudy to have any spectacular views, which was fine and I continued on my hike. I'm aiming to do 17 miles today and meet Fresh Ground at a parking lot. I ran into him while he was hiking the trail two days ago and we decided to meet at this location so that he can take me into town to resupply.

Day 124.

Today was the most physically difficult day I've had on the AT so far. I hiked over 18 miles and summited the peak of Mount Washington and completed all but one of the Presidential Range.

I did this because of the weather. The Whites are notorious for having some very bad, unpredictable weather, and by watching the weather reports carefully, yesterday was a beautifully clear day and I wanted to cover that ground while the conditions were good. Although, I did get sunburned on this long day. Some bad weather is headed our way and I did not want to be stuck in any storms or in any huts for a few days. Despite the difficulty, it was an amazing day. The views are incredible and well worth the hike. It was 12.5 miles of a climb straight up to get to the peak of Mount Washington. It was so exciting to get up to the top and see the surrounding mountains. I had an even better surprise when Fresh Ground was up there to feed the hikers.

He fed a whole bunch of hikers as we all filtered through during the day. I was starving so it was a nice surprise. After taking a lunch break and getting a few photos, I headed down to the Madison Hut. At the Madison Hut, I did my first work for stay. They allow thru-hikers to do some chores and work in exchange for us to be able to sleep on the floor in the dining room hall. It was nice to be out of the elements because it became very cold last night. After they serve dinner to the guest and ate dinner themselves, they offered dinner to me which I graciously accepted and ate as much as my belly could hold, which included three pieces of delicious chocolate cake!

Chapter 17: No Rain, No Maine. Almost there...

Trail lesson: *Surround yourself with the people that you strive to be and admire. Being on this trail and spending so much time alone has allowed me to reflect heavily about the people I surround myself with in my life. I believe that I am a stronger and better person by the people that I associate with. It has not always been easy, but I have significantly reduced or eliminated people in my life that are negative, draining, drama-filled, or just plain make me feel bad. Associating with more positive and determined people has helped me become more so that way myself, and I hope I have for them.*

Day 127. July 7, 2018

Yesterday I was hiking along a difficult section of the trail when I came across two female hikers that were so incredibly positive and just purely happy to be out there. They were taking a lunch break and offered me some trail mix which was very good. We ended up chatting for a little while before I went on my way.

Come to find out, I would be spending the night with them that evening.

I made it to the Carter Hut around 3:45 pm and did not want to continue on to the next shelter, over 7 miles away, due to the storm that was coming. I asked the staff if I could do a work-for-stay and they allowed it, even though I was at the facility earlier than they preferred hikers to be. I did not want to be in the way so I ended up going to one of the cabins and hanging out in one of the rooms, to stay out of the staff's way and out of the rain. About 30 minutes later, in comes the two ladies I met earlier that day. Kind of ironic that I ended up hanging out in the same room that they were assigned to out of the many rooms that they had available. We all ended up having a great conversation for quite a while before they said that I was welcome to stay in this cabin with them if the staff allowed it. There were many other bunks available in this cabin and they said that I was welcome to one of them. I didn't know if the staff would allow this, but I did go and ask and they said that would be fine. What a treat! I was able to sleep on the bunk bed mattress out of the rain, but the best part was it was like a slumber party with the two ladies. After doing some work for the staff and having a delicious dinner, I slept like a baby last night, warm and dry and in great company. It was a lot of fun. Their energy was just overwhelmingly positive and that was a huge uplifting surprise for me. I had not been sad or upset, just enjoyed the moment of hanging out with them. We ended up exchanging each other's information, so we can keep in touch, and I look forward to that.

Day 128.

I'm enjoying my hike as I'm walking along and as usual, throughout my day thinking about my husband, Matt. I was thinking about a communication breakthrough that we've had on this trail and thought it would be a good idea to share it. Early on in the trip, he asked me how I was feeling. I thought that was a very nice question to ask and then he asked again a few days later. After some thought, I realized that by him asking me that question it made me feel emotionally connected to him because I felt like he cared about my emotional state, which is important to a woman. I communicated with him that I appreciated him asking that question and since then he asks it just about every single time we communicate, which is almost every day. After being on the trail for so long I've received that question, many, many times and every time it makes me feel warm and fuzzy inside because it shows me that he really cares about how I'm feeling. I share this because I think it's a simple act of kindness that Matt does which makes me feel good. He would not have known this if I had not communicated with him that I appreciated that question. Communication is so important in a relationship and it makes me very happy when Matt and I are able to do something that brings us closer together. One more thought on this subject, I believe that asking how I'm doing and how I'm feeling is very different. When he asked me how I am feeling I feel much more emotionally connected than the general "how am I doing." Being more specific like that makes me feel closer to him. Maybe others can think about this and apply it to their relationship as well. It has worked for us.

Day 129.

Today was a long day but I am happy I made it through. I was able to get through the hardest mile on the Appalachian Trail known as the Mahoosuc Notch. It is 1.2 miles of large boulders where you have to put your trekking polls away and climb over them. At two points, you have to take your pack off and push it through small rock openings. I planned to get through this section of the trail while it was not raining and knowing that the next day will be rain, I had to push through it today. I felt great when I was done and happy that I did not have any significant issues getting through that mile. It took me an hour and 45 minutes to get through it as I took my time and was careful. What a relief when it was done!

Day 130.

I knew I needed to get to town today to resupply because I did not have enough food to make it to the next town that was 30 miles away. I looked on the map and saw that there was a road crossing that would be able to allow me to get a hitch and get to town for a re-supply. The notes on my Appalachian Trail app (App: Guthook, which can be used without Wi-Fi and is the best AT app to use for information and as a map.) from my phone stated that it is a road that is not well traveled, and one person even commented that he only saw one car in an hour. That had me a little concerned, but I needed to get more food one way or another, so I was prepared to wait if I had to. As soon as I made it to the road I said a little prayer asking that somebody would come along that was safe and would be able to take me to town for the resupply. I'm not even kidding, five minutes later a truck pulled up with two very nice gentlemen that stopped when I waved them down. I chatted with them for a moment to make sure that they were safe and then they offered to take me to town since that's where they were going. Their names were Steve and Matt. They were very nice and offered me a ride back after getting the supplies I needed. This was great since the town was about 25 minutes away from where I was. Everything went smoothly and I was able to get what I needed and get back to the trail in a short amount of time. After it was over I thought about what happened and said a little prayer to thank the higher power for sending me someone safe to get me to and from the trail and in a very timely manner. It was fun to chat with Matt and Steve along the way and I thanked them for their kindness before I continued on the trail.

Day 131.

Last night I was able to witness the most amazing thunder and lightning storm I've ever seen in my whole life. Thank goodness I was in a shelter. Around 2 o'clock in the morning, this humongous storm came in quickly and the rain came in sheets. The whole sky was lit up constantly by the lightning, and the thunder was so loud and close that I thought at any moment it would strike a tree and start a fire. Luckily, the rain was coming down so hard that I doubted a fire could really get started. I was so happy to be in the shelter because I had considered cowboy camping that night with a few other hikers that had decided to go down the hill to a campsite. I was just about to go and join them when another hiker told me that the bugs were actual bad down there since there was a lake close by to that location. I sure was thanking him at 2 o'clock in the morning when I wasn't getting washed away down the mountain during that huge storm. I tried to take a video of the storm but it did not do justice to how amazing and beautiful it really was.

Day 132.

Maine has been a bit of a challenge so far. For the past few days, there has been more mud in Maine than there ever was in Vermont. It's been a challenge hiking through mud and having wet feet all day, every day. I haven't been sleeping very well in the shelters lately because of people snoring or the excessive bugs that bite me all night long. Last night the bugs were a little bit tamer and amazingly everyone in the shelter did not snore, so I had some decent sleep and was so tired that I actually went to bed and fell asleep at 6:30 pm. What a world of a difference some decent sleep makes! I was up early and started hiking around 5:30 am. Starting early in hopes that it will stay cooler and the bugs won't be as bad for the majority of the day. I have a little over 200 miles to go and am anxious to get closer to the end. 132 days is a long time to be away from home and I am ready to get to the end.

I met some entertaining hikers last night and had a fun conversation before I went to sleep. Two thru-hikers and two flip-floppers were in the shelter and it was fun to compare hiking stories and talk about our gear and strategies. I love how everybody is out here working hard and going through the same challenges and are able to relate to one another. It feels like we are all in it together which is nice when you are hiking all day by yourself.

Day 136.

Yesterday I did my 14.5 strenuous hike for the day and then treated myself to a hostel stay at "Home of Maine" hostel in Stratton Maine. I was on the fence about staying at a hostel but had a hard time getting a hitch into Stratton to resupply, when the hostel owner showed up to pick up another hiker. I have not showered in a long time and the appeal of sleeping indoors and not getting eaten alive by bugs as I slept was very appealing. I took the opportunity and ended up staying and I'm glad I did. It was great to get everything cleaned and re-supplied, as well as a good nights' sleep. I also have not had cell service at all in Maine so it was good to connect with everybody and let them know that I was alright.

I spoke with my friend Richard and we planned on meeting at the base of Katahdin on July 14 and then summit together, seeing the mountain for the final finish the following day on the 15th. (ended up summiting a couple days early). That means I have to average 18.8 miles a day for the next 10 days, which I will be happy to do. I had heard the bugs were bad in Maine, but they really are relentless and bite me all day long. My body is covered in welts, and I will be happy to sleep indoors consistently very soon. The next two days should be some tough hiking with some higher peaks that I have to cover, but then after that, I think it will level out for a bit, which I'm looking forward to.

If you remember, I sent home my tent and cook system right before the Whites to cut down on pack weight, and have been shelter-hopping since then. With the southbound hiker bubble in full effect now, it is getting more and more difficult to get a spot at a shelter. That means I have to arrive very early to be sure I have a spot. I've been waking up at 4 o'clock in the morning and starting my hike around 5 am, which I don't mind, but it still does not guarantee a spot at my destination each day. Because of this, I purchased an $8 tarp yesterday and some rope which I will use to make a small little tent and sleep outside the shelters. I'm looking forward to trying that out tonight and think it'll work just fine until the end of my trek.

Emotionally, I am feeling good, it was really nice to talk to Matt yesterday and that always uplifts my spirits. I have very much enjoyed this adventure but am really looking forward to getting home. Knowing that I only have 11 more days of hiking is very exciting. I have learned so much and evolved a significant amount, more than I had ever expected and for that, I am extremely and forever grateful. It is a lot of hard work though. Hiking anywhere from 15 to 25 miles a day every day starts to wear on you emotionally, physically and everything else. I have not taken a zero day since I visited my family in Pennsylvania which was over 45 days ago. I'm not saying that to complain, I'm just trying to put into perspective the level of challenge this trail provides.

A lot of the southbound hikers start at Mount Katahdin in June and they are starting to catch up to where I am now. It is fun seeing so many more hikers on the trail and at the shelters which makes it more fun to converse with at the end of the day. I have to admit though, I look at their situation and see that they have 2000 miles more to go and I have less than 200 and feel almost sorry for them. They have such a challenge ahead of them, and I try to provide advice when they asked for it. Some of them seem out of their element and are still trying to adjust to trail life, just like I did in the beginning. I have come a long way since then and I know they will as well. I do see certain traits in some people and know that they probably will not make it to the end but some of them have that determination as I did in the beginning and that's fun to see too. I'm just so glad that I'm almost done and can't imagine starting the trail now in this heat and bug infestation. Although the snow was quite the challenge for a long time, I'm glad I started when and where I did.

Later that day… I was going to do a big mile day today but the temperature reached 91° and with all the elevation gain I had to climb, I decided to stop at the shelter called "Little Bigelow Lean-To." The shelter was especially appealing because it was right next to a water source they referred to as "The Tub." It's really a waterfall with a bunch of little pools that you can wade in and soak your feet. After I had set up the new tarp that I purchased yesterday, I went over to the tubs to soak my feet and wash my legs off from all the mud that was caked on them. It was the most refreshing and relaxing moment I can remember. It was so peaceful just soaking my feet in these pools of water next to a beautiful waterfall that I didn't want to leave, so I just stayed there until my stomach started growling and I had to go get some dinner. It was fun to just relax and watch the birds and squirrels play around. And the waterfall was so peaceful and beautiful, I wish this very spot was closer to my home because I would visit often.

The tarp tent I set up seems to be workable. I'm actually laying in it now as I am typing up these notes and it's quite small and cozy. I think it will serve the purpose and be exactly what I need to get me to the end of the trail. Talk about minimalistic camping, doesn't get much simpler than this, but I like it.

Day 137.

The tarp worked out well last night. There's not a lot of space but luckily, I'm not a huge person so it all worked out. The bugs wouldn't leave me alone last night and I was wide-awake at 3 am. Since I couldn't sleep, I decided to make a decision on my day today. I had two options: I could hike 17.5 miles to a shelter, spend the night there, and then hike the remaining miles to a river crossing where you have to take a ferry to cross the river. It's a dangerous river to cross and hikers have drowned so they make it mandatory that all hikers have to use the ferry. The "ferry" is actually guy navigating a canoe from 9 am to 2 pm. My second option was to get up and start hiking the 21.3 miles to the river by 2 o'clock. That was going to be a challenge but I decided to go with option B. I packed everything up, had my headlamp out, and was all ready to go, but could not for the life of me find the trail in the dark. After searching for 30 minutes I finally decided to go back to the shelter and wait for daylight. Some of the hikers in the shelter thought it was funny that I could not find the trail because they saw me go by with my light on early in the morning. After the sun started coming up, I was finally able to find my way back to the trail. It was going to be a strenuous push to get there in time but I did make it. After stopping for lunch at a hostel that was next to the trail, I continued on my way to the shelter for a 27.4-mile day. Quite proud of myself for doing that in 92° weather on less than four hours of sleep. Getting ready for those races.

I would like to explain the bug situation out here. A friend of mine jokingly said, "How bad can those bugs really be?" I know he did not mean any negativity from it, but I would like to tell you that they are really, very bad. It's crazy how many different kinds of flies are out here! Little gnats to big horse flies. I am covered in toxic bug spray chemicals and it only fends them off for a minute, literally. They fly up my nose and eyes, bite me through my clothes, and even bite me all night long while I'm trying to sleep. Therefore, sleep is impossible. The bugs out here suck! I have even considered hiking huge mile days to be done and get out of this bug infested state. Thank goodness there's a thunderstorm coming tomorrow because I would love to hike in the rain to avoid these bugs any day. They are a constant huge annoyance that never stops. I would say that the bugs on the Appalachian Trail are the most annoying and awful part of this trek. I'm not just having a tantrum, it's bad out here for everybody, and they say the same thing. Just teaches me another lesson, I will never take for granted going to sleep indoors without being eaten alive again. I was optimistic at the beginning of the summer and thought, oh I'm not going to use toxic chemicals out there. I'll just use my face bug net and everything will be fine. Boy, was I wrong in every way: the bugs still get to your face with a bug net and bite the heck out of you. Especially when you're asleep and the bug net is against your skin, it's as if you're not even wearing it at all. The only option is to cover up in your sleeping bag from head to toe and practically have a heat stroke at night trying to get a few minutes of peace. I have more bug bite welts all over my body than I have ever had in my life. A lot of them are now open sores because they itch constantly and I have scratched them until they have become sores. As of today, I have exactly 144 miles to go, which I plan to do in nine days. I'm very fortunate and happy to be almost done with this trail and feel very, very sorry for all the south bounders that started in June that will have to endure the wrath of these bugs and heat this summer. I can see how someone can really go crazy out here because of them. Now all of that snow in March doesn't seem so bad, LOL.

Later that day… About 45 minutes after I wrote that last paragraph about the bugs in my notes, I was pleasantly surprised to run into a very nice trail angel. There was a gentleman by a parking lot waiting for three female hikers that he was there to pick up. He started a conversation with me and asked me about my hike and how much further I had to go. During the course of our chat, he offered me a Gatorade and then gave me a pack of peanut M&Ms and a snickers bar. That was all I needed to turn my whole day around. Sometimes the kindness of these trail angels, when you're having a particularly low point of the trail, is extremely helpful.

A few miles later I ended up getting to the shelter I was going to stay at for the night. About a half a mile from the shelter there was a lake that hikers had written in the log book at the shelter, was a good place to go for a swim. I had not gone swimming yet on this adventure so I grabbed my washcloth and went in search of this very nice lake. It turned out to be absolutely incredible.

The water was a little chilly but after standing in it for a few minutes you adjust to it. I was able to rinse off and wash my stinky sweaty body and rinse out my hair. I felt like a new woman afterward. The lake was so beautiful and had people that lived on it that were out on their boats enjoying the hot day. I sat on the dock for a while and just soaked in the beauty. Again, I did not want to leave but my stomach had other plans and demanded dinner. I went back to the shelter, had a big dinner and then promptly fell asleep. I slept so well since it was a very hot and hard day on the trail.

Day 138.

I slept in this morning and had a late start hiking. It was already raining outside so I wasn't in a big hurry to leave. Yesterday I prayed for rain and today I enjoyed it. I'm thrilled because it is almost noon and I have not had one bug bite or the threat of heat stroke yet today. It's a great day! I did not even bother wearing a rain jacket and I am soaked to the bone but I am a happy camper.

Two hours later the rain stopped, the sun came out, and the bugs resumed eating thru-hikers with a vengeance. It was nice while it lasted but now they are making up for lost time.

Chapter 18: Mt. Katahdin in Maine, I did it!

Trail lesson: *I started this Appalachian Trail adventure as a scared girl. By the end, I discovered my strengths and became a courageous woman.*

Final Day on the Trail. July 13, 2018

Today I finished the Appalachian Trail! 2,190 miles, 145 days from Georgia to Maine, and from February 18 to July 13. I have been overcome with emotion since I made it to the finish. I had so many things I wanted to say in my video speech at the end, but as soon as I hit record I just went blank and became overwhelmed with emotion. I had finished and was so proud and happy as tears of joy flew. I'm not really one to cry very often but this occasion was out of my control. As difficult and challenging as the Appalachian Trail is, I could not help but cry tears of joy as I walked to the final finish sign, placed my hands on it and said: *"I did it."*

It took me 145 days to walk from Georgia to Maine from February 18 to July 13. That's a long time to be away from home but there is no way to put a price on the experience and adventure I gained. All the incredible people I met were also kind beyond words. The beautiful scenery I was able to personally witness is indescribably nature's beauty. All the lessons I learned along the way that I hope to never forget. One of the things that happened that is most meaningful is that I learned how strong and brave I could be. Situations and times where I had become so afraid or uncertain, I learned how much inner strength I really have. The inner strength we all have is much stronger than we could ever imagine, as it was for me I discovered. We all are a lot stronger than we ever give ourselves credit for. My body is stronger than I ever imagined but my mind is the powerhouse behind it all. Once you set your mind to something and give it your whole heart, I think you can accomplish just about anything. (Safely and within reason of course).

This has been a life-changing adventure for me, for the better of course. It was very difficult but I don't regret a single second of it and cherish not only the amazing times but the hard ones as well. I learned so much from the whole experience which has changed the course of my life.

I was very fortunate to have a good friend of mine, Richard, up there on top of Mount Katahdin with me as I finished. He has been one of the most supportive friends along this whole journey and I am forever grateful to him. He surprised me by coming up to Maine early and picking me up yesterday. After a great dinner and a good night's sleep, along with a shower, we tackled Mount Katahdin together today. Richard, thank you so much for all your endless support and care. You helped me along this journey immensely and I cannot thank you enough. Thank you for being there with me at the end to share that very special moment.

As I finished my speech for my video and had a few pictures at the end with the sign, several people came up to congratulate me. One of those people was a thru-hiker that finished a month ago and just wanted to come up to summit Katahdin again. He gave me a high-five and congratulated me. Another girl came up to me who was starting her Appalachian Trail journey today from the very spot that I was standing. She saw me finish and said that I was very inspiring to her and then asked if I had any advice for her. I told her a few things and wished her the best of luck on her journey as she starts it today.

As I walk down Mount Katahdin right now I just cannot believe that I did it. Still flooded with the emotions and I am still processing everything from the whole experience. I'm walking along and my mind wanders to all the unique and amazing people I met, all the shelters I stayed in or near, the States I walked through, the funny things that happened, as well as the scary ones. All the rain and snow that I went through, but just as important, the amazing peaks I was fortunate enough to summit. I am incredibly grateful and fortunate to have had this experience and will cherish it for the rest of my life!

If I could say anything to any of you out there that you take away from this experience, it would be to dream big and then go after it with your heart. The goals that are difficult and really challenge you as a person are the ones that are most incredible and worth it. So much had happened on the five months that I was on this trail, that I could never have planned or imagined. I do have some big goals that I will be tackling in my near future, but I will always cherish this adventure for the rest of my life.

Chapter 19: A log of my daily hikes, mileage breakdown and places I stayed each night....

Trail lesson: *Less is better unless it's love or gratitude....or water.*

Daily mileage:

Day 1: Approach Trail 7.6 miles, 7.4 miles done to Hawk mountain campsite.
Day 2: 16.5 miles to Lance Creek camp site.
Day 3: 13.9 miles to Whitley Gap Shelter (alone)
Day 4: 12.1 miles to Blue Mountain Shelter (total 49.9)
Day 5: 15.5 miles to Deep Gap Shelter (elevation gain today 4569).
Day 6: 12.6 miles to Bly Gap Campground (town day at Hiawassee)
Day 7: 15.3 miles to Carter Gap Shelter
Day 8: 12.1 miles to Rock Gap Shelter
Day 9: 3.6 miles to Winding Stair Gap (went to Franklin. Stayed at a Hostel)
Day 10: 15.8 miles to Cold Spring Shelter
Day 11: 18.6 miles to Sassafras Gap Shelter

Day 12: 15.2 miles to Cable Gap Shelter (mile 158.8, right before the Smokies).

Day 13: 7.1 miles to Fontana Dam Visitor Center

Day 14: 17.4 miles to Spencer Field Shelter

Day 15: 19.2 miles to Mt. Collins Shelter

Day 16: 20.6 miles to Tri-Corner Know Shelter

Day 17: 18.4 miles to Standing Bear Hostel (green corner rd. Mile 240.3).

Day 18: 15.2 miles to Roaring Fork Shelter

Day 19: 18.2 miles to Hot Springs (Laughing Heart Hostel).

Day 20: 14.8 miles to Allen Gap (in-laws picked me up).

Day 21: Zero Day (first one).

Day 22: 12.1 miles to Jerry Cabin Shelter (hiked with co-worker Bryan).

Day 23: 15.5 miles to Hog Back Ridge Shelter.

Day 24: 2.4 miles to Sam's Gap (Trail angel Eric took me to his home to stay out of the snow storms).

Day 25: Zero-day (snow storm).

Day 26: 18.3 miles to No Business Knob Shelter

Day 27: 23.3 miles to Cherry Gap Shelter

Day 28: 3 miles to Iron Mountain Gap (hitched to Erwin).

Day 29: 9.4 miles to Hugh's Gap Rd. (Fresh Ground food and campsite)

Day 30: 12 miles to Overmountain Shelter (converted Barn).

Day 31: 7.6 miles to HWY 19. (Roan TN at Mountain Harbor Hostel).

Day 32: 26.6 miles to Laurel Fork Shelter

Day 33: 30.2 miles to Double Springs Shelter

Day 34: 19.3 miles to Damascus VA. (Lazy Fox Inn).

Day 35: Zero-day (Lazy Fox Inn).

Day 36: 0 miles on AT. Did 12 miles of the Virginia Creeper Trail (Stealth camped under the Creeper Trail Bridge on the AT. Made up the miles in a few days on this section.)

Day 37: 19.3 miles to Wise Shelter

Day 38: 7.3 miles to campground and mile 512 with "Fresh Ground."

Day 39: 33.1 miles to Fox Creek (mile marker 512/Fresh Ground location) from US route 11 near "The Barn" Restaurant (mile marker 545.1)…was driven 33.1 miles out so we could hike back to Fresh Ground and meet my Mom Friday. (Longest Hike so far).

Day 40: Zero Day (with Mom, Brandon, and Millie)

Day 41: 13.6 miles backward (sobo) from the VA Luther Hassinger Memorial Bridge (where the AT and Creeper Trail meet). Had to make up this part since I hiked the Creeper Trail out of Damascus.

Day 42: Zero Day (with Mom, Brandon, and Millie).

Day 43: 14.7 miles from US Route 11 to Knot Maul Branch Shelter

Day 44: 24.5 miles to Laurel Creek Campsite (mile 584.4).

Day 45: 25.3 miles to Trent Grocery (stayed overnight.

Day 46: 17.5 miles to Docs Knob Shelter

Day 47: 8.3 miles to Parisburg, VA.

Day 48: 7.8 miles to Rice Field Shelter

Day 49: 25.3 miles to War Spur Shelter

Day 50: 18.2 miles to Niday Shelter

Day 51: 16.8 miles to VA route 624.

Day 52: 16.3 miles to Lamberts Meadow Shelter

Day 53: 14.4 miles to Fullhardt Knob Shelter

Day 54: 16.6 miles to Bearwollow Gap (Family picked me up there to spent the weekend.)

Day 55: Zero Day (Day with family). (Hiked 8 miles to McAfee Knob with my in-laws).

Day 56: 10.4 miles to Bryant Ridge Shelter

Day 57: 22.6 miles to Matt's Creek Shelter

Day 58: 22.2 miles to Brown Mountain Creek Shelter

Day 59: 8.1 miles to Hog Camp Gap

Day 60: 21.9 miles to Harpers Creek Shelter

Day 61: 22 miles to Paul Wolfe Shelter

Day 62: 10 miles to Bear Den Camp Ground (mile 868 Stealth camp).

Day 63: 28.9 miles to Pinefield Hut

Day 64: 21 miles to Bearfence Mountain Hut

Day 65: 7.1 miles to Wayside Restaurant-Big Meadows (went to Luray for the night because of weather)

Day 66: Zero Day (Race rest)

Day 67: Zero Day (Race rest)
Day 68: Zero Day (Fly to CA)
Day 69: Zero Day (Race rest)
Day 70: Zero Day (Race Day)
Day 71: Zero Day (Race Recovery)
Day 72: Zero Day (Fly back to VA)
Day 73: (back on the trail) 15.3 miles to Byrd's Nest Hut #3
Day 74: 17.5 miles to Gravel Springs Hut
Day 75: 24.1 miles to Manassas Gap Shelter
Day 76: 19.8 miles to Sam Moore Shelter
Day 77: 2.9 miles to Bears Den Hostel
Day 78: 16.2 miles to "4-mile" campground
Day 79: 21.3 miles to Dahlgen Backpack Camp Site
Day 80: 24.2 miles to Falls Creek Camp Site
Day 81: 20 miles to Quarry Gap Shelter
Day 82: 24.4 miles to Tagg Run Camp Site (water source)
Day 83: 12 miles to Boiling Springs PA (TCO outfitter) Matt
picked me up for a visit.
Day 84: Zero Day
Day 85: Zero Day
Day 86: 14.3 miles to Darlington Shelter
Day 87: 22.3 miles to Peters Mountain Shelter
Day 88: 18 miles to a tent site. (Mile 1177.0)
Day 89: 23.1 miles to Hertline Campsite
Day 90: 18.3 miles to Port Clinton Hotel
Day 91: 22.5 miles to Allentown Shelter
Day 92: 17.3 miles to Parking lot at PA 873 (family Sue and
David picked me up and took me to their home)
Day 93: Zero Day with family
Day 94: 16.2 miles to Leroy Smith Shelter (then hiked 1.5 miles
to Sue and David's house for the night)
Day 95: 24.9 miles to Backpackers Campsite
Day 96: 20 miles to Brink Road Shelter
Day 97: 19.6 miles to High Point Shelter
Day 98: 12.4 miles to Pochuck Mountain. Shelter
Day 99: 18.6 miles to a Stealth Camp area
Day 100: 15.3 miles to Elk Parking Lot (mile 1385.3) Friend
Richard picked me up and I stayed in a hotel this night.

Day 101: 14.7 miles to Stealth camping area near a viewpoint.
Day 102: 21.7 miles to a Stealth Camping site.
Day 103: 25.2 miles to Telephone Pioneer Shelter
Day 104: 21.2 miles to Mt. Algo Shelter
Day 105: 22.7 miles to Belter's Campsite
Day 106: 14.5 miles to Brassie Brook Shelter (food coma after Fresh Ground visit).
Day 107: 22.8 miles to Tom Leonard Shelter
Day 108: 21.3 miles to Upper Goose Pond Cabin
Day 109: 20.4 miles to Levardi's (campsite in the back of a hiker's home).
Day 110: 20.7 miles to Wilbur Clearing Shelter
Day 111: 23 miles to Melville Nauheim Shelter
Day 112: 20 miles to a Stealth Camp spot next to a creek. (Black Brook).
Day 113: 21.5 miles to Bromley Mtn. Cabin (free stay for hikers).
Day 114: 26.7 miles to Minerva Hinchey Shelter
Day 115: 22.4 miles to Inn at Long Trail (stealth camped at a nearby side Trail after I had dinner at the Inn.
Day 116: 15.4 miles to "Lookout Cabin"
Day 117: 27 miles to Stealth Campsite (mile 1746.3)
Day 118: 29.6 miles to Jacobs Brook (stealth camp at mile 1775.9)
Day 119: 11.2 to Road Crossing NH Route 25C mile 1787.1 (Richard picked me up for resupply and sleep, returned me to the same location).
Day 120: 12.9 miles to Beaver Brook Shelter (First day in the NH White Mountains)
Day 121: 13 miles to Kinsman Pond Shelter
Day 122: 15 miles to Garfield Ridge Shelter
Day 123: 17.6 miles to Parking lot US Route 302 (Stealth Camped with Fresh Ground and other hikers.)
Day 124: 18.2 miles to Madison Spring Hut (Work for stay)
Day 125: 7.8 miles to Pinkham Notch Visitors Center (Slept in Fresh Grounds Car)
Day 126: 5.9 miles to Carter Notch Hut (Work for stay)
Day 127: 13.3 miles to Rattle River Shelter

Day 128: 13.7 miles to Gentian Pond Shelter
Day 129: 14.7 miles to Speck Pond Shelter
Day 130: 10.3 miles to Frye Notch Lean -To
Day 131: 10.5 miles to Hall Mountain Lean-To
Day 132: 12.3 miles to Bemis Mtn. Lean-To
Day 133: 19.5 miles to Piazza Rock Lean-To
Day 134: 16.8 miles to Spaulding Mountain Lean-To
Day 135: 13.5 miles to Parking area at mile 2002.7 (stayed at Hostel of Maine for the night).
Day 136: 15.3 miles to Little Bigelow Lean-To
Day 137: 27.4 miles to Pleasant Pond Lean-To
Day 138: 22 miles to Horseshoe Canyon Lean-To
Day 139: 15.8 miles to Stealth Campsite by a waterfall. (Mile 2083.2).
Day 140: 19.1 miles to Chairback Gap Lean-To
Day 141: 17.1 miles to Logan Brook Lean-To
Day 142: 23.1 miles to Potaywadjo Lean-To
Day 143: 18.2 miles to Rainbow Stream Lean-To
Day 144: 24.9 miles to Katahdin Camp Ground (Richard picked me up).
Day 145: 4.9 miles to Mt. Katahdin….THE FINISH!!

Chapter 20: What I learned while on the Appalachian Trail....

Throughout the hike, I would realize different concepts and have a lot of "ah ha" moments. They started happening more frequently so I began to write them down as they happened. Mainly, so I wouldn't forget them. Here is a compilation of what I learned and realized while on my Appalachian Trail Thru-Hike.

-The weight of items in your pack is determined more so by how much it all weighs soaking wet.

-Cheap gear does not usually cut it on the AT. A good tent, sleeping bag, pack, and rain jacket are essential when you start in February.

-It can snow and rain at the same time. I know this first hand.

-I have met the kindest people on the trail. The AT experience gives a whole new faith in humanity and the level of kindness people can have.

-The Trail does provide. It provides the opportunity to work out whatever emotions you need to with help along the way from amazing people through an abundance of opportunities.

-Always, and I mean always have your rain gear accessible. Even if it says perfect weather, it will rain. Be ready for rain at all times unless you are prepared to be wet. Don't forget that everything wet weighs a lot more.

-Completing a through hike of the Appalachian Trail, running an ultra-race, and living life, in general, is accomplished with what is between your ears. Focus on what you're doing and stick to your goals and you can do just about anything.

-What people think about me is none of my business. So, don't worry about petty drama, it's a waste of energy and attention.

-People typically tend to see what they feel. Meaning, if they are having a bad day, everything they encounter may be negative. If they are having a good day even bad situations have a positive Outlook. Keep that in mind when people are trying to perceive and understand you.

-Life is full of big and small opportunities, be sure to have enough time to take them. It's nice to hike fast and do big mileage but it's also important to not miss the iconic scenes and places to go. I'm glad I was fortunate enough to see so much on the trail and off.

-It is much easier to hike with fewer things. Much less to carry makes the hike more enjoyable and easier. I believe this could be applied to life as well. The less stuff we have the easier and more efficient life can be.

-On the good days and bad days, you still take one step at a time the same way. The good days will fade and the bad days will too, but just keep moving forward and enjoy the journey.

-Things will all fall into place if you just let it happen. The less I try to control a situation and the more I focus on going with the flow, the easier things turned out to be.

-There will be moments when you are exhausted, wet, cold, or really hungry, but embrace the feelings and know that it won't last forever. Just like with difficult situations in life, it will pass as well.

-Listening to other people is more important than proving you are right. I strive to listen more and argue my case less.

-Happiness is experienced every single day. It's the type of life you choose to live. Happiness is not acquired by reaching a goal or destination; it's the journey of life. Everyone is in control of their own journey.

-I realize now that if you look to other people or person to be happy you will eventually always be disappointed. Happiness comes within oneself and is lived every day.

-If you're struggling to have more inner peace, then make alone time a priority. Having so much alone time out here has allowed me to reach a level of inner peace I never fathomed as possible. I don't think someone needs to spend five months hiking alone to achieve inner peace but making alone time a priority would help.

-It's ok to express emotions and cry if you need to. Don't hold it in.

-It is all about the journey, so when you get the chance to listen to an elder talk about their past, sit down and listen. The little old German woman I met on the trail in Waynesboro VA taught me a lot with what she said and didn't say. I saw how kindness is lived through the life of this old woman. Enjoy the journey, slow down, and most of all….listen.

-Nature can help you heal if you let it.

-I've come to realize that if you truly love someone, you want them to be happy no matter what. Happy doing what they love and feel they need to do. I feel I have this kind of love with Matt. He has never once put a limit on me nor do I on him, we both want each other to do what makes us happy. Sometimes that is together and sometimes that is with me being in the woods for 5 months or running long distance races. Our love is strong whether we are physically together or not, distance doesn't change our connection to each other, nor should it. I love you, Matt.

-BLS before ALS. What I mean is referring to something that we use in the medical field. Basic life support before advanced life support. I think this can be applied to all aspects of life. I believe that you have to have a solid foundation in the basics before you can advance into more complicated situation and skills. For example, be a basic good human being and then from there advance on to doing more good deeds for humanity and to help others, but you have to start being good and true to yourself before you can help others. Basics first.

-One of the reasons I had to come out to the trail is to shake out my raincoat. Let me explain. The raincoat is a representation of my life and the areas of my life. The water droplets on top of the raincoat are a representation of things that I was giving time and attention to that were draining me of energy or were not useful to me. I had to shake out my raincoat in order to focus less on society, focus less on my ego, focus less on social media, focus less on drama that is no good for me, focus less on negative people, focus less on food and activities that were not healthy for me, and focus more on things that were important or are important in my life. Important things like my family, my values, spirituality, healthy foods and habits, a healthier frame of mind, better sleep, more positivity, and the things that benefit me in my life as I move away from things that don't. I discovered that I came out here to "shake out my raincoat."

-Observe and acknowledge your emotions, you're having them for a reason. Suppressing them can be dangerous. Look at emotions as guidance rather than a negative thing that should be suppressed.

-Passion. I believe everyone is passionate about something. If you don't know what it is then you should figure it out. While out on the trail I asked a bunch of people what they were passionate about. The people that were doing something towards their passion for what they are passionate about tended, to be happier people in general. I think there's something to it. Just to name two, I am passionate about running and spirituality. The more I have embraced the two of them and enjoyed it, the happier I have become.

-Everything that happens in our lives, the good and bad things, are opportunities to learn and grow. Everything happens for a reason, and we are given opportunities to develop every day. It's our choice to learn from them or not.

-I started this Appalachian Trail adventure as a scared girl. By the end, I discovered my strengths and became a courageous woman.

-The love, honesty, and positivity that I have created in our relationship with Matt have fueled my strength while on the AT. We have created a beautiful and healthy relationship that we both have wanted. It has not always been easy but it is based on values that we both consider important. It's a beautiful relationship for us, and it makes me very grateful, happy and most of all, loved.

-Sleep. I am sleeping better out here on the trail than I ever have in my whole life.....when the intense snoring is not an issue of course. I thought long and hard about why that is the case and finally figured it out. I worry significantly less than I ever have in my adult life at this point. Having less worry and stress allows me to sleep soundly. I fall asleep right away, stay asleep all night, and wake up rejuvenated with an abundance of energy. No melatonin or any sleep aids needed. Better and sounder sleep is a great side effect of less worry and stress in my life.

-Surround yourself with the people that you strive to be and admire. Being on this trail and spending so much time alone has allowed me to reflect heavily about the people I surround myself with in my life. I believe that I am a stronger and better person by the people that I associate with. It has not always been easy, but I have significantly reduced or eliminated people in my life that are negative, draining, drama-filled, or just plain make me feel bad. Associating with more positive and determined people have helped me become more so that way myself, and I hope I have for them.

-Make personal time for yourself a priority. I realized that to change up the monotony of walking all day in the woods I would look forward to a movie date with myself. I had downloaded onto my phone a bunch of movies from Netflix and once or twice a week I would make it a point to get to my campsite a little early and have time to enjoy a movie before I went to sleep. I know it sounds silly but it was some fun "me time" that I looked forward to and made me happy. I know people may think that I should not bring movies and technology into my trail experience but it was something that made me happy, that I enjoyed and looked forward to. Once I realized that I made it more of a priority. I think everyone should find something that they enjoy and make it a priority to enjoy it on a regular basis.

-A big lesson I realized after thinking long and hard is how I perceive life. I can perceive it in a positive way or negative way and I will give an example. I struggled a lot with the breakup of my ex-boyfriend recently. For a long time, I perceived it as a negative thing that happened and a great loss. Again, having so much time out here to think about things, I eventually realized I needed to change my perception of the situation. As soon as I started perceiving that relationship as something that taught me so much, provided a lot of good times, and remembered all the negative things that I did not have to put up with or endure any more, my attitude about the whole situation changed. My point is, you can perceive every situation in your life in a positive manner or a negative manner. Even better, perceive it as a learning opportunity and figure out what the lesson was and that may help you move on as it did for me.

-Controlled Perception. We can choose how we perceive situations in our life, of course. We can choose to perceive things in a positive manner or negative manner. Our life is how we perceive it and it is our choice and how we want to do that. For example, all of the elevation gain challenges with climbing hills/mountains that I have to get over had me thinking. I can look at it in a negative way and say, great, another hill that's going to be hard to get over and will make me tired. Or, I can change my perception and think of it as, great, another hill that I will climb that will make my muscles stronger and more capable to handle other hills and the races that are coming up. Each hill makes my body stronger and my mind as well. It is all in how I choose to perceive it.

-Comparing yourself to others can be harmful. Being out on this trail, there are women that are significantly stronger hikers than I am. A few times I've caught myself comparing my abilities to theirs. After some thought, I realized that it is not only pointless but can be harmful. We are all out here for different reasons and those reasons motivate us to have stronger days or days that we need to take a break and rest our bodies. Everyone says "hike your own hike" but until I really started putting that into practice and not comparing myself to other hikers, I struggled a little bit. Once I stopped comparing myself to other hikers and focused more on what I was doing, I had a significantly more positive and stronger experience. I intend to try not to compare myself to others after the trail as well. We never know where someone is in their life so how can we compare ourselves to them or even judge them for that matter. I will strive to not compare or judge others, especially as easily as I used to.

-Nature and the earth are completely beautiful. Being out in nature for the past five months has granted me a new appreciation for its beauty. Because of this, I have a new disgust for littering. Please do not litter anywhere. Take it a step further and pick up litter that you see and try to recycle. When I get home, I intend to make preserving this beautiful earth more of a priority.

-Many times, I would be walking along with my eyes looking down and my head down, focusing on the path. I would do this until I realize that my neck was stiff and then would look to my left, then to my right. Once I started looking around me more often my neck was much less stiff and I would get to see and enjoy nature and the wildlife that surrounded me much more. I think I need to do this in my own life after the trail as well. It's easy to get tunnel vision on the day to day tasks and forget to take a moment to look around to notice the beauty of our life that surrounds us. I will make it a point to be more present in day to day life and look around so that I can miss much less of what is going on around me.

-Less is better unless it's love or gratitude....or water.

-When you get to a technical part of the trail; slow down, breathe, and take it one step at a time. When you get into a challenging part of life; slow down, breathe, and take it one step at a time.

-I don't learn very much when things are easy, but I learned a HECK of a lot when things are hard. Embrace the difficulty and appreciate the opportunities.

-When you have love and support at home (my husband), you have incredible strength within you and it doesn't matter where you are.

-Forgiveness. After having so much time to think out here, I realized that dwelling on things that happened in my past that were caused by people had a negative effect. Holding onto those memories and constantly thinking about how bad I feel about them is not healthy. I realized that I needed to come to terms with exactly how I felt about the situations. I realized that people and situations come into my life to teach me something, whether it be good or bad situations. The bad situations do tend to be toxic, especially when they were held inside. I received tremendous relief and peace when I confronted the emotions and then forgave the person that caused them. People can release those emotions in many different ways. For me, I wrote a letter to the person that caused a lot of negative emotions for me and then sent it to that person. I think forgiveness is important, but even more important is the lessons we learned from the situations that we went through. Everything happens for reason and purpose, and when I look at things that way I'm no longer a victim.

-We all have great potential, more than we know. So, think big and know you can do it. When I started the Appalachian Trail, I didn't know how I would get through it, but I did and it has changed my life for the better in many, many ways. Think big and go do it, because we all have unlimited potential.

-Embracing my path and doing what I feel I need to do in my life regardless of what other people think or express what I should do. I believe that we are all unique and develop in our own unique ways. To embrace this, we need to follow our own unique path. For me, it is being out here on this trail where I am getting a huge amount of opportunity for self-discovery. For me, it is also doing long-distance races where I discover so much about myself, how I want to live my life, and so much more. I believe that people need to embrace their unique ways and path regardless of what other people think they should do. It was suggested to me that after my first hundred-mile race I completed, that I should stop. I'm so glad I did not stop running long distance because it has led to so many amazing opportunities. It was expressed and implied that I should not do this trail, but I did what I felt I needed to do, and it has been life-changing for me. I hope that everyone embraces what they feel they need to do, regardless of what society or people close to them suggest or tell them they need to do. We need to live our own lives and create our own path!

-One of the things I am most grateful for is my relationship with Matt. Our relationship is based on love, kindness, and honesty. With a base of those virtues, we both have the strength to do anything.

-It's great when it's raining on the trail because the bugs leave you alone and don't annoy you during the hike. It's great when it's not raining on the trail because you don't hike the whole day cold and wet. There's good and bad to every situation, we get to choose which side of the coin we want to focus on.

-There will be times when people want to start a conflict with you. This is the perfect opportunity to focus on how you want to handle the situation. It is hard for someone to argue with you if you do not argue back. While in such a situation recently, I chose and hope to continue to choose, to not respond with negativity and to send that person love instead. Most of the time the situation will be resolved on its own with time. No need to give your attention and focus to negativity when you don't have to.

- I have realized that it is important in life to try to be more present and enjoy the moment. About halfway through the hike, I found myself stopping to take in the sights, a feeling, an emotion, and experience. I realized in those moments that I am exactly where I need to be at that moment. By doing that, it made me feel good, and I began to do it more often. The more often I did it, the more present I became, which made me less worried about the past, less worried about the future, and more appreciative of exactly where I was and what I was doing at that exact moment.

-Thunder and lightning storms are fascinating to me, but thunderstorms on mountain tops are spectacular!

Chapter 21: Adjusting Back to Home Life and deciding on a Career Change

Trail lesson: *I've come to realize that if you truly love someone, you want them to be happy no matter what. Happy doing what they love and feel they need to do. I feel I have this kind of love with Matt. He has never once put a limit on me nor do I on him, we both want each other to do what makes us happy. Sometimes that is together and sometimes that is with me being in the woods for 5 months or running long distance races. Our love is strong whether we are physically together or not, distance doesn't change our connection to each other, nor should it.*

I was so grateful to so many people that were incredibly supportive of me during my thru-hike. I took a moment to write Fresh Ground a little message to express my gratitude towards all the support he gave to me and all of the hikers out there. Here is what was sent:

"Fresh Ground, I want to take a minute to thank you from the bottom of my heart for all the help you tirelessly provide for all of us on the trail. Your help and support go WAY beyond feeding very hungry hikers. I had the pleasure of seeing you so many times along the trail and each time my belly was filled with much-needed food, but my spirits were always lifted. You are a real angel. It was the stories you told, the fist pumps to all of us when we arrive, the endless food you gave (even at 10 pm when I arrived so hungry and you were already in your hammock going to sleep, you got right up and fired up the stove and made us so much food), all the trail advice that I really needed, the safe place to sleep near your camp area, charging my phone multiple times when I was nowhere near a town, bringing me Gatorade when my stomach was upset, waiting for me when I hiked slow and you already fed everyone else just so you could feed me too, surprising me with birthday visit and meal, boiling me hot water so I could clean up myself, letting me sleep in your car when it rained, picking me up on the trail so I could stay with everyone at camp, all the laughs and memories along the way. You do FAR more than feed smelly hungry hikers... you help to build us up along the way. I always looked forward to seeing you and felt so recharged as I hiked on. I've never seen someone work so hard to serve and help others from the kindness of their heart like I have witnessed you do so. You are always there for us and I thank you immensely for all your help and support. You are THE AT TRAIL ANGEL."

After I completed my thru-hike of the Appalachian Trail, Richard was kind enough to drive me to his home in Massachusetts. I spent a day and a half there with him and his family before flying back home to Charlotte NC. I could not wait to get home and see Matt and Millie! I spent the next few weeks adjusting back to home life and trying to recover from the trail as I prepared for the 200-mile races. I wish I had a little more time at home before the first race but I still had a solid 3 weeks before they would begin.

One action I made a priority to do before I drove across the country for the Bigfoot race was to de-clutter all the junk out of our house. We had so much stuff we didn't use or need. It felt great to get rid of a lot of unnecessary things that just took up space. Literally carrying everything I needed on my back for the last 5 months proved to me that I really didn't need a lot of the junk I had at home.

Another decision I had when I made it home was not to return to my position as a Paramedic. Matt and I decided that I would take this year off to hike and do the races so there wasn't any pressure for me to return to work as soon I made it home. One of the reason's I went on the trail was to contemplate a career change and figure out if that was a direction I should take since I already had in mind something completely different to try. Well, after a lot of thought and a few signs along the way, I decided to go for the new career direction with all my energy and effort. So what is it, I'm sure you're wondering.

Throughout my life, I have been somewhat intuitive in a metaphysical sense. I very much so believe in God, angels, spirit guides and the fact that they can communicate with us while we are having an experience here on Earth. There have been times in my life when I would receive somewhat of a "download" of information about a particular person or situation that turned out to be useful for that person or that would give me a better understanding of the situation. A few weeks before going on the trail I had an experience where I actually channeled my higher self and was able to communicate with a collective group of guides and angels. This blew my mind and I soon became fascinated with this new-found way of communicating. I had been doing this to an extent all my life but this particular instance showed me what was happening. I would mentally throw out a question and I would receive an answer in various ways. I soon discovered I was clairaudient, clairvoyant, clairsentient, able to ask about past lives, communicate with souls that were living or passed on, as well as other abilities.

It has been an interesting experience to be able to communicate with my higher self and team for the benefit of helping people. They have taught us so much about the universe, religion, and really anything we ask about. Working as a paramedic for the past 9 years, I have enjoyed the ability to help people in need. I have found that by communicating in this way, I have been able to have a much greater impact on people and really help them to understand things in their lives and in many cases heal from them. While on the trail I thought about the abilities a lot. I'm confident that this was my path in life.

Here is my website if you are interested in learning more about what I'm working on in regards to this: stephaniechannelsspirit.com

In order to communicate in this way, I have to be in a high vibration which allows me to come close to their vibration, or frequency rather. In order to do this, I have to live a low-stress life so I can concentrate on communicating with them once I am able to connect with their energy. They love to use metaphors to communicate a point and have given us some great analogies to many of our questions.

I have been getting used to the way they communicate with me, which is through energy. I may not have been able to pick up Spanish very well in college, but the language of energy in other dimensions seems to be my forte. With that being said, I am focusing on living a low stress, simple, drama free, high vibration life so that I may continue to channel. Each session allows me to get more confident, understand the energy better, find out new abilities, and understand that I am on the right path.

Chapter 22: Big Foot 206.5 Mile Race, August 11-14

August 16, 2018

What a race! I don't even know where to begin so I guess I'll start with a little pre-race travel. Before I get into the race details, I would like to say that this was the most difficult race I have ever done in my life. So many highs and lows during the event which were all very significant.

The race took place in Washington State which is completely across the country from Charlotte North Carolina where I live. Plane and hotel expenses were quite expensive so I had to resort to traveling by car by myself. It was a 41-hour drive and I made it in 3 1/2 days while sleeping in rest areas in my car along the way. It worked out well and I was quite comfortable as I made my way across the country. The United States is so beautiful to see and I was fortunate enough to drive through it and soak in the sights as I traveled.

Thursday before the race, I went to the pre-race meeting to pick up my bib and swag. They had a mandatory meeting where they discuss the course and some of the potential problem's runners could face. As soon as the meeting was over at 4 o'clock, ate dinner and went straight to the campsite, where I was able to park my car and sleep. I went to bed by eight and was asleep by nine.

Race day; I didn't sleep that well the night before the race because I was worried I would miss the multiple alarms that I set up. There is a time change as well, so in the back of my mind, I was worried that the time would get screwed up somehow. I was awake before my 4 am alarm and really didn't have much to do to get ready. All the runners met at the high school by 5 am to load the buses which left at 5:30 to take us to the starting line, which was 2 hours away. I tried to rest while on our bus ride but did not get any sleep. It was nice to rest and relax though before the race. I was a little nervous about this race because I had heard of its difficulty level. I knew I was as ready as I would ever be but I didn't feel like my body was recovered enough from the Appalachian Trail since I had only finished it 3 weeks prior to this race. We arrived at the starting line in plenty of time to stretch, apply sunscreen, have a little snack, and take a picture.

I would be running for the next four days straight. Little scary! I was off to a good start and had a very slow easy pace. There were so many people but I knew we would all be scattered out after a while. A few miles into the race we had to go over all of these scrambling rocks which reminded me of Pennsylvania on the AT. Then we had to use a rope to climb up a very steep part of a cliff which was actually kind of fun. Then we were all in the sun for the next 40 miles. It became so hot with temperatures in the low 90s and the sun just beating down on us. People were throwing up and one guy said he was peeing blood. I wasn't feeling too good myself, even though I had slowed my pace and was trying to take it easy. That heat was just killing us out there. By mile 18.1 which was the Blue Lake aid station, I knew that this was going to be a lot more difficult of a race than I had imagined. My strength was being zapped by the minute as we were climbing these brutal inclines in this extreme heat. I was drinking plenty of water but my appetite was not cooperating. The views were spectacular though. We ran around Mount Saint Helen which was a volcano that had irrupted in 1980. Then we went over lava fields which were cool also.

As I made it to the Windy Ridge aid station I was feeling awful. Between the heat and not being able to eat too much, I was really not feeling well. I was hoping that cooler temperature at night would help me feel better, but I just wasn't finding my strength. This was the first time that I was considering the possibility that I may not be able to finish this race. I arrived at the Windy Ridge aid station and considered throwing in the towel and thought that I could either quit and go sleep in my car or stay in the race and sleep on the chair next to a heater at the aid station. I choose the heater and cozied up in a big chair with a warm blanket and slept for 2 1/2 hours. This was definitely not the plan to do this in the first 15 hours of the race but I could not help how exhausted I felt, and decided to take the time to rest since I was ahead of the time cut off for that aid station. After my 2 1/2-hour nap I was actually feeling better and was able to eat something. I decided to go to the next station and put on an audiobook as I took off. This was a beautiful section of trails as I watched the sunrise through the mountains and saw a little deer family with the mother deer, buck, and their two little babies.

I made it through the Johnston Ridge and Coldwater aid station without any issues other than my legs already feeling stiff and sore, and my left knee giving me some issues with stiffness. I knew it was about to start raining soon so I packed my rain gear before I left this aid station.

The 18.7 miles to the Norway pass aid station was the most difficult part of the entire race for me. This is where I hit a wall that lasted almost the entire 18 miles of this section. This was a particularly hard part of the race due to the elevation gain/loss that was steep, along with rain. Towards the second half of this section, my left knee started locking up, and I was having a really hard time bending it. Every step was excruciating and I was starting to do 30 to 35-minute miles. Going downhill was the worst. This is where I hit my one and only wall but it was a bad one. I felt like my body was giving out and I had no confidence that it could go another 140 miles once I reached the next aid station. I did not know what to do. I was really extremely upset, to tell you the truth. Thoughts going through my head were that I hiked the entire Appalachian Trail in preparation for these races, and here I am with my body giving out. I was telling my body to go forward and move, but it just wouldn't do it. I did not know what to do at this point and I was just so devastated about having to quit the race. I had driven all that way, 41 hours driving to get to this race. I felt bad because I thought I was letting Matt down because, I wanted to take this year to hike the trail and do these races, so how could I quit. Then I thought about this book that I'm writing and how the whole book is based on the Appalachian Trail and the completion of these three races, how could I write this after quitting the first race. I also thought about how the scenery of this race was. So beautiful, and I really wanted to see the rest of the race. How was I going to do that if I quit? I wanted so desperately for my body to cooperate but I felt like between not recovering enough from the Appalachian Trail, not getting the best sleep in my car before the race, too long of a drive to get there, major heat we had to deal with on the first day, I just felt like my body had reached its limit of what it could do. I felt worn out and completely defeated. During my break down I cried in the mountains as I hobbled along and prayed for someone or something to help me get my body to cooperate so that I can finish this race, which I so desperately wanted to do. I had thought about this race and prepared for it for so long, and I wanted to finish it more than anything.

I arrived at the Norway Pass aid station in so much pain! They asked me if I was alright, and I said that my left knee is giving me major trouble and that I was going to get something to eat and see if I felt better. Another runner was there that had a lot of experience and asked me where my pain was. The pain was on the left side of my left knee cap and was excruciating. He said that he thought it was an IT band issue. Whatever the issue was I was not able to bend my knee hardly at all. After getting something to eat and resting for a while I tried to see if there was any improvement and there wasn't. I made the decision that I was going to have to DNF (DNF: did not finish/quit the race), for my first race ever. I hobbled over to the aid station volunteer that was keeping track of all the runners and with tears welling in my eyes, I told him that I was going to have to DNF. He asked me if I was sure that's what I wanted to do and I said absolutely not but I can't seem to bend my knee and didn't think that I could go on. He said okay, and told me that I would have to wait four hours for a ride back to the start line after the aid station closed. I did not have anybody with me and no other choice so I said that if I had to wait for four hours that if you could please take me off the DNF list and that I would try to stretch my leg and try to get it to bend. He agreed and then immediately grabbed a mat, laid me on my side and massaged my IT band area for 20 minutes. Then he showed me a stretch I had never seen before, that allowed my IT band to really open up. That stretch saved my race! Just as I was stretching, a sweet lady came over with two Motrin and gave them to me. People were so nice to me. After 30 minutes I was able to bend my knee and felt I could keep going. The two volunteers that came to help me were so kind and happy to see me continue. The woman gave me a hug and the gentleman gave me well wishes as I took off to the next aid station which was 11.1 (my bib number was 111) miles away.

I was so thrilled to be able to actually move and I did a light jog for most of this section of the race. I was so happy! I really did not want to quit this race and without those two volunteers to help, I believe I would have. I made it to the Elk Pass aid station without any issues. There were wild blueberries along this route and I ate handfuls of them on my way. So delicious! Once I made it to the aid station I had a little snack and since I had plenty of time, decided to lay down for an hour and a half. This was not a sleep station but a volunteer offered to let me lay down in her car. I snuggled into the back of the car and shut my eyes for an hour and a half, dozing on and off but not really falling asleep for any significant period of time. It was really nice to be off my feet though. At this aid station, I met a really cool French couple who I would run on and off with throughout the race. They had become really cold so they were trying to warm up and rest at this aid station as well.

The next few miles went on without any issues. I was just thrilled to be able to move my legs and be back in the race. I had never come so close to quitting a race before and it really disturbed me. At the aid station called Road 9327, I decided to rest again. I lay down for an hour and a half and was actually able to get some sleep. At 3:30 I woke up and took off. I had a nice cushion of time before the cut off times but I wanted to maintain that. As I left this aid station the sun came up again on the third day of the race. This was the day I actually was able to settle down into a good rhythm and get into the Game. I was feeling so much better, was able to jog, and knew that I was going to finish the race so I really started to enjoy it. The first two days were REALLY rough but I had made it through them. The next two days would not be quite as bad, despite exhaustion.

I started seeing familiar faces from time to time and it was fun to chat with runners at the aid stations. We were all struggling out there and we were all trying to be as supportive as we could when we had the opportunity. One runner gave me some herbal cream to put on my knees at one point in hopes to help my pain, which it really did. Another runner saw that I was struggling with nausea early on in the race and gave me a pill he had called Zofran which helps with nausea. Huge help! It was really nice of him to do that and after I took the pill, an hour later I was able to eat a hamburger. That helped me feel a lot better. That's the thing with the ultra-running community; you meet the nicest people in the world out there. We all want to help each other get to the end and if there's a way we can do that we always do. Even if that's a piece of gear we could offer or some words of encouragement, everybody is out there to help, almost 100% of the time. I am honored to be a part of this community.

From there, I made it to the Lewis River aid station and a bunch of other runners were discussing how difficult the next section of the trail would be. It was almost 20 miles long but the terrain and elevation were going to be very difficult. It was 2 pm and I thought it would be a good idea to take a little rest before I tackle that section. I went into a sleeping tent to lay down for an hour and a half but sat there worried about getting through that section and having to do a lot of it in the dark. Since I wasn't able to relax I decided to go ahead and get up and get moving. I'm glad I did because it was a really tough section. I made it to the Council Bluff aid station at 10:30 pm totally exhausted. As I arrived a volunteer approached me and asked if I needed anything and I told him that I had to lie down for a moment before I could continue. He started giving me attitude and telling me that it was not a sleep station. I told him I understand that and I would be happy to lie down on the ground. He was very grouchy and rather rude to me and said that he could not guarantee I could have a blanket for very long. I told him that I was so tired and could not continue the next 10 miles to the next aid station without lying down for a few minutes and I will be just fine lying on the ground with my jacket. I went over to get some water and another runner came to my rescue with him telling the volunteer that he did not have to treat me that way and be so rude to me when I was clearly exhausted. I really appreciated him doing that and thanked him multiple times later. I did go lay down in a little patch of grass for 20 minutes before another volunteer came over and told me to get up and get moving. I thought this was crazy because I wasn't bothering anybody and I clearly wasn't welcomed at that aid station, a big contrast from every other aid station. I was so disturbed that I collected my gear and took off. The next 10 miles were not easy either and I had to lie down on the side of the trail twice. I would've rested longer but I was so cold and would wake up shivering so I had to keep moving since I couldn't sleep being so cold. Temps were in the low 40's. I finally made it after four hours to the Chain of Lakes aid station and was so ready to crawl in a tent to take a nap. The volunteers at the station were much more kind. I told them that I had to lie down and they gave me a blanket and directed me to a tent. I took a hard, three-hour nap which was very much needed. I woke up feeling so much better and ready to tackle the next section. It was still dark when I woke up, which I like to do so that I can enjoy the sunrise, it seems to rejuvenate me and it did. I started running with two other runners I had met and we had a nice conversation that morning. The sunrise was spectacular too! During this section, we had to go through a lot of water crossings, so the ice-cold water woke us all up as well. It was kind of funny watching all the runners struggle through the water while I really enjoyed it actually, especially going through so much water on the Appalachian trail. I didn't think anything of it.

I made it through two more long sections of the trail which took all of the daylight time. The sun went down and I had my sights set on the twin sisters aid station because that's where I was going to have my next rest. I met up with the two runners I had run with earlier that morning and they were starting to fade on energy. They asked if I would like to join them for the rest of the distance to twin sisters and I agreed to. I wasn't feeling too bad so I tried to do the talking and keep them engaged. The male runner in front of me was starting to drift off the trail from time to time and the female behind me was having some crazy hallucinations (dead animals, body parts, and a piano). We eventually made it to Twin Sisters and all agreed to take a nap. I had lots of time so I set my alarm for four hours. Of course, I was not able to sleep much at all because first off, my hips started cramping so I knew I was dehydrated, so I took an electrolyte supplement and ibuprofen, along with more water. Then I settled down into the cot again when multiple car alarms started going off and car lights shining brightly into our tent. It was as bright as a football stadium at night with the spotlight on us inside the tent. Then two other runners came into the tent and were trying to get settled. All in all, I didn't get a whole lot of sleep and decided to just get up and keep going after three hours. It was enough for me to keep pushing. The climb out of twin sisters is steep so I was ready to knock that out.

The sunrise was beautiful again and would be the last one I would see on this trail. I had till 6 pm to finish this race and was doing very well on time so I just took my time and enjoyed the last day out there. The next two aid stations were very nice and I took my time having a snack and enjoying the conversation with all the other runners. The volunteers were excited for us because they knew that we would be finishing, at least most of us. Some did not make the cut off time, unfortunately. I did get really tired at two points and had to lie down on the side of the trail for a quick cat nap before making it to the end. At about 8 miles to the end of the race, I finally had cell service and decided to call Matt. I was so happy to speak with him and hear his voice. I told him about my experience and things that had happened and that kept me alert for a while. It was a nice treat to talk to him and that lifted my spirits a lot.

I made it to the high school and did my loop around the track into the finish line. Many of the runners I had met along the way were at the end and cheered me on. It was such a great moment and I was so excited and happy to be finished with my first 200-mile race. I really didn't think I could do it at that one point and I was so happy that I pushed through it and made it!

After finishing the race, I was able to pick out my own custom-made belt buckle. Then I sat in a chair and just enjoyed the moment with all the other runners for quite a while after a little nap. They had great food for all of us so we all hung out and cheered on the remaining runners. The last two runners to come in were so inspiring since they struggled so much to get to the end that there was hardly a dry eye out there as they crossed the finish line.

That is how the first of the three races went, quite an experience for sure. I thought back on what I did in preparation for this race. I do think that hiking the Appalachian Trail helped prepare me for the difficulty of this race. I knew there would be a lot of elevation climbing and I felt strong enough to handle it. There were a lot of technical parts of the trail which I also felt ready for. The only thing that I did not consider was the lack of rest. On the Appalachian Trail, I would do very difficult back to back days but I was always able to rest at night in preparation for the next day. For this race, we were doing back to back difficult days with little to no sleep. Sleep deprivation really gets to you as time goes on, especially on a 4 1/2-day race! Towards the end, we were all laying down on the side of the trail for small naps and that seemed to work very well. I had done that in the past but not as many times as I did on this race, and will definitely be implementing this in the future. It's just enough rest to keep you going.

I also felt confident doing this race on my own. At one point, at an aid station, a volunteer asked me where my crew was and I told him I didn't have one. He was shocked that I was doing this race without anyone with me, No pacer either! I told him it was just too far for people to travel but that I was doing fine out here. I do believe the Appalachian Trail prepared me for being alone out there for long periods of time. I didn't have any issues with it at all, and at no point was I scared of anything in the woods. I'm grateful that I was able to hike that trail in preparation for these races and feel it did do a lot of good.

Another very important thing that I believe worked in my favor was my mindset. Being out on the Appalachian Trail I worked through so many emotions and issues that I had. I didn't know how necessary that would be. When you are doing long-distance races like this, you hit walls and the first thing to come up is anything that you are possibly upset about. I really wasn't upset about anything and I feel I'm in a really good place in my life, so when I did hit low points they weren't aggravated by negative emotions and feelings. That has happened to me in the past which has been a struggle to get through. I am fortunate to have worked through so many things and be at such peace. I think that was very important in this race.

It took a while for it to sink in that I accomplished this amazing race. It is the most difficult one of the three and people were saying that to get through this one puts the runners on a graduate level as a runner. I would have to agree with them since the difficulty of this race was quite intense. I am so happy and thrilled that I made it through, and did not quit. If those volunteers had not been there to help, I don't know what I would've done. That's another big lesson learned, even when you think you don't have any options if you just slow things down you can usually find a solution. A bunch of the runners I met out there are also doing the Triple Crown (all three 200-mile distance races this year). We exchanged some information and said that we would all look forward to seeing each other at the next race. That's going to be a lot of fun. I'm looking forward to it already. Now I need to just recover and rest up for my next Tahoe 200 race in three weeks. To get home from the Bigfoot race, I drove to Salt Lake City. I visited with my good friend Alicia and then flew back to Charlotte, leaving my car at her house until I returned for the next race.

Chapter 23: Tahoe 200 Race Report, Sept 7-11

September 7-11, 2018.

The Tahoe 205.5 mile race was a success! This race went much smoother than the last Bigfoot race I did three weeks ago. Here's the story of how it went down.

I knew that the elevation of this race was going to be the biggest challenge since the entire race takes place between 6000 and 10,000 feet. Living in Charlotte North Carolina at 600 feet, I am not used to that type of altitude. For this reason, I made it a point to try to acclimate to the elevation for the race as much as possible. Prior to the race, I flew to Salt Lake City to spend a couple of days with my good friend Alicia. She lives at 4000 feet, so I spent one night with her, and then spent another night at a nearby mountain at 6400 feet. Then the next day I went to Reno, Nevada to spend an evening with my family that lives there. They live at elevation as well and it was great to visit them after not seeing them in many years. Then the night before the race I spent car camping at the race start which was at 6000 feet. The day before the race we had the usual pre-race preparations. We had to do a medical check, get our race swag, our picture was taken, and attend the mandatory meeting. The most fun of the day was seeing a lot of the runners I had met at the previous race, that was also doing the Triple Crown. Chatting and hanging out with them the whole day was a lot of fun. I was able to see Chris, who was the guy that helped me with my IT band issue during Bigfoot and really saved my race. I was so happy to thank him in person for all his help. After getting everything ready for the race and dropping off all of my drop bags, I went to the car to relax until bedtime. I went to bed early and had a really good night's rest.

Race day finally arrived, and the race officially started at 9 am on Friday morning. I wished all my running buddies' good luck and headed towards the starting line. As I found my place at the start of the race, I looked over and saw a female runner that I highly admire! I saw Courtney Dauwalter getting ready to start the race as well. She is the big up-and-coming ultra-runner right now and holds many course records. Her ability and talent are phenomenal and it was really cool to be in the company of such an elite athlete. I could not resist the urge and had to be a fan, so I pulled out my camera and walked right up to her and asked to take a picture with her 30 seconds before the start of the race she allowed me to get a selfie with her which made me so happy. She ended up having a very good race and finished the race over 10 hours in front of the current course record, amazing!

The race was off to a good start but I knew it was going to be a tough day. We had a lot of climbing to do that day and the temperatures were rising rapidly. I took it very slow and steady to try to conserve energy, and stay as cool as possible. Not too long after the start, people were already having issues with the altitude, so I consciously took it a little slower. I could push it later on in the race to make up some time, but I did not want to start the race in bad shape. Early on in the race, I met many nice runners, and we chatted along the way, especially when we would be near the Lake Tahoe water. Very nice to meet new people and hear their stories. One runner showed me a tattoo that his first young child hand drew that he had put on his arm for inspiration. I the tattoo was unique and sweet.

The first day and night of the race went smoothly. I had my music to listen to, and a few documentaries I downloaded to listen to as well. I made it to the Brockway summit aid station which was a sleep station and had planned to take a little nap at the 50-mile point. As I approached the aid station I was pleasantly surprised to see my Uncle Mike there to greet me. He had come to visit and show support, it was a great surprise. He brought me two turkey sandwiches, two big Twix bars, and a banana. Very nice of him to do that and I was thrilled to have the food since aid station food is not always appetizing. After restocking my pack, I went to lie down for a 1 1/2-hour nap. Uncle Mike waited until I woke up and saw me off. The next section of the trail was probably my most favorite part of the whole trail. It was a big climb out of that aid station, which then crossed over the mountain ridges with the most spectacular views of Lake Tahoe. It was absolutely gorgeous!

Day 2: Smooth progress as I was keeping up on my calories and hydrating well and just settling into the race. I was definitely noticing the altitude and feeling its effects. The higher I climbed in elevation, the more it felt like I was walking through quicksand. I noticed it would take one and a half to two times the effort to move as I normally would so that was quite a challenge. I was still going well, and I did not have any major aches or pains. A blister was forming on my right big toe and another toe on my right foot, but that is typical. At one point I decided to change my socks and as I peeled off one of my socks, the toenail on my right foot came off with the sock. The blister on that toe was so bad that it had already detached. I pulled the remaining part of it off and pulled on some new socks. Luckily it did not hurt at all and I was able to keep going without a problem. I later learned that it takes about 6 months for a new big toenail to grow back if you are ever wondering about that.

Night 2: I ended up at the "Heavenly" aid station which was a sleep station. The sleep station was interesting because it was at a ski resort so they had it set up for the runners to sleep inside. It was warm and full of blowup mattresses and cots. After a quick snack, I headed in there to lay down, my alarm set for four hours. I had built up a decent buffer of time and was able to afford the time for a nap. I settled into my blowup mattress and had a solid three hours of sleep before a runner came in and started snoring so loudly that he literally cleared out the room. Everybody got up and started taking off, including me. I bet it was funny when he woke up since he arrived in a full room of people and then woke up to an empty room, lol. I had a solid three hours of sleep before that happened. I felt good with that rest and ready to tackle day three. Day three had the highest altitude points of the race, which was a challenge. I knew it was coming and kept my pace conservative so that I would not get to a point where I started feeling sick. Luckily that worked out very well. Another day of amazing views to enjoy!

A running friend of mine named Leslie was at this race. I ran with her for quite a while during the Bigfoot race, and she had returned to the Tahoe race to crew for some runners and help out as a volunteer at the stations. She was the one with the crazy hallucinations during the Bigfoot race. She went to each aid station and volunteered as the race went on. With all the work she was doing, she looked just as exhausted and tired as the rest of us runners. I give her a lot of credit for all the hard work she did during this race because she had even less sleep than we did. At each aid station I saw her, she made it a point to give me a hug and compliment before I left. So sweet!

Night 3: I made it to the "Sierra Lodge" aid station which was at another ski resort that they had set up for us to use during the race. This was another great spot because it was indoors also. Again, after getting something to eat I went to go lay down. I went into a sleeping room and there was only one blow up mattress available, so I claimed it. All the lights were out so I had to use my phone light to navigate to my spot. As I approached the mattress, I noticed the guy sleeping next to me was completely naked! He was not even wearing socks. I found this quite funny and kind of giggled, hoping that he was comfortable. He did not bother me whatsoever, and I quickly fell right to sleep for a solid four hours. I was feeling really good and refreshed after that rest and left the aid station ready to tackle day four.

All along the race, I was trying to maintain a nice buffer of a few hours before cut off times at each aid station. This worked out rather well, and I knew that I could rest at any point if I needed to. Not having that pressure allowed me to really enjoy the race a lot more. I was thrilled that besides some blisters on my toes, I was feeling pretty good. Amazingly, I never hit a wall during this race, a big contrast from the last race. People asked if this race was more difficult than the last one and I think in a way it was because of the altitude difficulty. Just the extra added effort to move made it more difficult to get the miles done. Even though Bigfoot had more climbing/elevation gain, this race required a lot more effort, so I voted for Tahoe as more difficult. As needed, I did take 15-20 minute cat naps along the trail. This helps so much when you are exhausted and need a rest to be able to continue. I feel I have mastered napping along the trailside efficiently.

Day 4: I had a smooth day on day 4 as well. I enjoyed the Tahoe race more because I liked the views and trails better than Bigfoot. Up to this point, the trails had not been too technical, but little did I know what was ahead of me. During the day it became quite hot and many runners were struggling with the heat. I slowed down to compensate for the higher temperatures, which worked out well. I was still able to eat and hydrate, so I was pleased. On day 4, I think I enjoyed the race the most, it was exciting to know without a doubt that I would be finishing the race without any problems. I had plenty of time and was maintaining a decent pace easily.

As usual, I did a lot of pondering about how fortunate I am to be out there doing this race. How fortunate I am for the time, my ability, the love, and support from so many people, and most of all, the love and support from my husband, Matt. We have a solid foundation of a marriage, and because of our strength together, it fuels my inner strength. My favorite thing to think about on the trails is how fortunate my life is, and this race was no different. It really makes a big difference when you are in a good place in your life while you're out there on a race. So many things can go wrong and challenge your mental ability to get through these races. When you have less stress and drama in your life, it significantly benefits your mindset out on the trails. I've noticed this in the last few races, actually since being on the Appalachian Trail. My life is far less stressful and much more positive. This allows me to focus more on the good aspects of my life rather than negative parts of it. Of course, we all have stress in our lives but by reducing it as much as possible, it really makes life a lot smoother. I could not be more grateful for this. I thank the Appalachian Trail for helping me achieve a level of peace within myself and attribute the trail to the ability to tackle a lot of these climbs. Many times, during the race as I was marching up a particularly difficult section of the trail I would think, this was like that part of the trail on the AT. I hiked the Appalachian Trail to get my legs ready for these races and was definitely feeling the power that I needed during the race.

I listened to a lot of songs during the race, but there were two songs that I listen to repeatedly. "Limitless" by Ramses B, the words in this song really resonated with me out there. And "Road Less Traveled" by Lauren Alaina. Road Less Traveled because that pretty much sums up my life, LOL. Limitless because there are phrases in that song that I feel give me strength during the race.

I was moving right along on day 4 and had established enough time to get to the "Loon Lake" aid station for a two-hour nap before tackling the next section. Once I arrived at the aid station the medic looked at me and asked if I needed anything. I said no, that I was feeling good but wanted to take a little nap before heading on. She suggested that since I looked good and not in any distress that I continue going. I sat there and thought about it for a few moments and decided to have a snack and to do just that. Little did I know that the next section of this trail was going to be the most difficult part of the whole race. It is known as the Rubicon Section and it is 17 GRUELING miles of 4 x 4 Jeep tracks. Large rocks covered in a very thin layer of fine dusty sand. The whole trail goes either straight up or straight down and was very slippery. I did this entire section in the dark making it even more difficult. Breathing in all that dust, and having a really hard time seeing was a big challenge! It felt like the section would never end, but eventually, I finally made it to the last station of the race known as "Bakers Pass". I was so happy to get there. I arrived at Bakers pass at 3 am with more than enough time to finish the last 7 miles before 1 pm the next day, so I decided to relax at that aid station for a little while. I had some hot soup from the volunteers and sat around a portable heater and had a conversation with the volunteers about the two top winners of the race. Everyone was really nice and it was fun to sit, relax, and chat with them. Then I started getting sleepy and decided to take a nap, which turned into a two-hour nap, before getting up and heading out. One of the volunteers was very nice and offered to give my feet some attention. We had a small water crossing during the last section so my feet were soaking wet and very cold. We took the wet shoes and socks off, and he very kindly scrubbed my feet down with wipes, getting all the many layers of dirt and dust off of them, which was quite a job. Honestly, as he worked on scrubbing my feet, it felt so incredibly good, I felt I was in heaven. After such great feet attention, I had a nice nap before getting up and heading out to accomplish the last portion of the race.

I left the last aid station right before the sunrise and was able to watch a beautiful sunrise on my last day of this race. It was gorgeous. There was a surprise at the finish line waiting for me, my Uncle Mike was there. It's such a great surprise to see someone you know that is supportive, especially on such a challenging race. He was there at the finish line, ready to take my picture and show his support. I was just overjoyed with happiness at the end.

At the end of the race, I remember thinking that I was feeling rather good and if I had to go on running that I felt like I could. I was surprised that throughout this whole race I did not really hit any walls at all. I wanted to remember this feeling so I have the confidence to tackle Moab in a month. The race is 240 miles with 112 hours to do it. I finished this race just over 95 hours and took it really nice and easy. I could've pushed it much harder but wanted to enjoy the journey way more than getting a fast time. I'm happy with that decision and really did have a good time on this race, which I feel for me is what it's all about.

After chatting with some other runners that finished, getting something to eat, and collecting/getting my drop bags to my car, I laid down for a nap. I took one of the hardest naps of my life for four hours. I had a small blow up mattress with covers set up in my car and I slept like a rock. I meant to see the last runner come in but I didn't wake up and missed it. Once I was up, I talked to friends before heading over to my uncle's house for dinner. They kindly let me stay there the night which allowed me to get cleaned up, have a nice dinner with them, and a good night sleep before hitting the road to drive back to Charlotte. It is a 38-hour drive back to Charlotte so I needed to rest up before tackling that next challenge.

Besides a big blister on big right toe, I amazingly did not have any injuries or discomfort. My IT band never bothered me, legs felt good and strong, my feet held out well. I did the whole race in one pair of Altra Loan Peak shoes with multiple sock changes. I think the altitude forced me to go a little bit slower at times when I normally would've pushed it, which helped not to break down my body. I stayed warm enough during the cold nights when the temperatures were in the 30s, and as cool as I could be when temperatures were in the upper 80s during the day. The drastic temperature change was interesting, but not bad to handle. I did not have a problem with eating and drinking, so that was nice too. I stayed positive throughout the whole race and remained in good spirits the whole time. Surprise visits from my Uncle Mike those three times contributed to my positive attitude as well, thank you, Uncle Mike. He was great support with sandwiches, Twix bars, even five-hour energy, and just him being there. I gave up caffeine over a month ago but plan to make exceptions for caffeine while I'm racing. I think that worked out to my advantage, because over the course of this race, I had a total of 1 1/2 5hr energy bottles, and I felt the effects strongly. Unfortunately, I could not resist the urge of caffeine and picked up the vice after returning home. Alcohol was way easier to give up than caffeine.

I also really enjoyed seeing all of the friends that I had made in the last race, at this one. I know I've said it before but it is so true, the Ultra Running community is absolutely amazing. We are all out there to support each other and every opportunity to do that is taken. I may be out there alone for a good part of the race, but I really feel like I'm surrounded by friends.

The long drive home is going well and I'm having a good time reflecting on some highlights of these last two races. Still can't believe I'm fortunate enough to be able to do this at this time in my life. The thought that I had while I was running was that I have had some intense factors during many of my races over the past three years of my Ultra-running journey. I thought I would share them.

Extreme heat-Key West race.
Extreme sand-Destin races.
Extreme Thunderstorms-many races but mainly Indiana trail.
Extreme cold weather-Tunnel Hill race.
Extreme elevation-Bigfoot 200 race.
Extreme altitude-Tahoe 200 race.

Extreme muddy conditions-Hallucination race the first time.

It just goes to show the vast majority of challenges that ultra-races can throw at you. I love this aspect of the race and embrace the challenges each one brings. It's like playing a game of Chess, you need a strategy to handle different situations. I think the ability to adapt to all types of conditions makes a good runner. I feel that these races have made me a better, so much growth from each race along with so many great memories.

Now it's time to rest and recover as I prepare for my last race of the season. Moab 240 in Utah

Chapter 24: Moab 240 Mile Race October 12-16, and the completion of the Triple Crown!

October 26, 2018. Moab 243, Triple Crown Completion.

It took me a few days to process everything that happened during this last race. It was the last race of the Triple Crown of 200's challenge and it was a heck of a race. 243 miles, extra miles because of a course change, through the desert, rocks, boulders, snow, and ice with temperatures at night in the single digits. It took me 107 hours to complete the race. I had 112 hours to do it so I was within the time allowed. Of the three races, Moab was by far the most emotional race for me. Mostly good emotions but the race did not start off that way. Here is how it all went down.

This race was unique compared to the other two because I was going to have a crew of one, my friend Richard, there to help support me. He was amazing support and sponsored this entire race for me! He paid for me to fly to Salt Lake City, where I met him since he flew in from Massachusetts. He rented a Suburban truck and we headed to Moab. Richard has been a family friend for many years and has always been my biggest support in these adventures. I knew it would be a lot of fun and very helpful to have him there to help me. The first night I went to bed very early and slept as much as possible to try and conserve all energy for the race. The next day was spent doing the pre-race check-in, runners meeting, medical check, getting last minute supplies, and then going to bed early. As always, it was so fun seeing my running friends that I have become to know over the last three months. I also checked in with a company called Kogalla who have created a running light system that is awesome. They gave me a light system to use/have on the trail which worked amazingly.

I was not nervous about this race until about an hour before the race started on race morning. I felt ready, rested, recovered from the other races, and happy to have a crew, but something in the back of my mind told me that this was going to be a difficult race. Turns out it was.

The race started at 7 am on Friday morning without a problem. We ran through the town of Moab and straight into the boulder filled red mountains. I was in the middle of the pack and we all started spreading out after a few miles. Once we ran into the canyons we had to rely on the markers that were set on our path to figure out which way to go. I had already gone the wrong way once and had to get back on the trail, so that wasn't too good of a start. That minor detour had me concerned on how well the course was marked from there on out. The course was also a little more difficult than I had anticipated at the beginning of the race. We had quite a bit of climbing to do right from the beginning. My body did not feel very strong and I was losing confidence as the miles went by. The first aid station that I was going to see Richard at was at mile 17.9. As I came into the aid station I was feeling rough and I had low confidence for the rest of this race. I was also slightly nauseous, which is never good for me during a race. I don't deal well with nausea. I also started off the race without trekking polls and that was throwing off my stride and balance big time. I was thrilled to see Richard with my polls and gear at the first aid station that crews were allowed to meet at. Thank goodness he had them and brought the rest of my gear because I used them for the rest of the race. I didn't think I would need them in the beginning because I thought the route would be flatter, but I was wrong. As rough as I was feeling, I knew I had to keep moving. I drank some Tailwind which is an electrolyte supplement mixed in water and was off. I wouldn't be able to see Richard again until mile 73.7 because they did not allow crew access at a single aid station until then.

I took it slow and just concentrated on one mile at a time till I felt better, which thankfully I did. I know I was getting overwhelmed by the course at the start of this race because of its difficulty. I also was having a rough time connecting with the trail. On the Appalachian Trail and the last two races, a large majority of it was in the woods. I feel a lot more comfortable in the woods and being able to connect to the trail with that sort of terrain. I know this may sound odd, but being out in the red clay and rock was very difficult to get into the race for some reason.

The first evening, as the sun went down, I was running with two groups of runners. I was going back-and-forth between them as my pace would increase and decrease. It was fun to listen to conversations they were having and to interact with them as we all made our way through the canyons. I was able to use my Kogalla light system for the first time and it was awesome. One runners' light was fading because of a weak battery but I was able to light up the whole path, which was great. I ran through the first night without any issues and finally made it to the aid station at mile 73.7 the next morning where I found Richard. I decided to take my first nap and slept in the warm truck for a solid two hours at the "Hamburger aid station."

The second day went smoothly, as I just focused on making it from one aid station to the next. I was starting to feel better and as I went, became stronger. We were able to run through some sections that had trees and bushes which made me feel more connected to the trail. I didn't realize that was so important to me until this race. Richard was working hard, volunteering at all the aid stations, while he was waiting for me to arrive. From the aid station at mile 73.3 on, he met me at every single aid station until the end of the race. That involved a lot of driving and navigating. This was quite a challenge for him because the directions to the aid stations were not always clear and he had to do some off-road driving to get to some of them which were quite a challenge. In all, he drove 750 miles while I was running, over the 5-day race period. It was very helpful to see him at each aid station though and I more often than not tried to get a little nap in the car before tackling the next section.

Along the path, I would occasionally run into one of the photographers named Scott. He was always so positive and happy to see each of the runners and I looked forward to seeing him. I had met him at each of the races since he was one of the two photographers that were on the courses for each race. To be honest, I developed a little crush on him. I made it a point to try to chat with him a little bit in each section where we met.

Going into the evenings of the third night we ran into quite a bit of open fields with some strong winds. I had plenty of layers of warm clothes so I didn't mind it at all. I was moving along strong and as I said earlier, as I went I seemed to get stronger. I was really able to get into this race on the third and fourth day and started to enjoy it once I knew that I was going to make it. Once I was through the initial first few aid stations I was good to go. Having a nice warm car with an air mattress to sleep on when I was able to get some sleep was wonderful. I know they had sleep stations with heated tents, but to have some quiet alone time made a big difference in my physical strength during this race. Again, I was so grateful for Richard and everything he did!

As always, I did a lot of thinking during this race. I couldn't believe that this was the last piece of the puzzle to an incredible year of travel and adventure. I had worked and planed so hard to get these goals accomplished and I was finally on the last leg of the challenge. I did a lot of reflecting on what I have been through this past year. It honestly blows me away that I have been able to do so much, and safely. These three races motivated me to hike the entire Appalachian Trail, and in that journey, I was able to do some remarkable self-discovery. I'm so grateful for all that I have been able to do, see, experience, and the time to discover so much about myself. The help and encouragement from friends, family, and strangers have been so incredibly touching. I ran through the last half of this race filled with great memories and reflection on what had happened. A part of me was excited to finish this race but another part of me wanted to keep going so it wouldn't end.

I still had half the race to go and it sure wasn't getting any easier. On Sunday, the temperatures overnight dropped into the single digits with the wind chill, making it even colder. The AT prepared me for harsh weather, and I was fine with the conditions throughout the whole race. I know one runner was sent to the hospital for hypothermia. My staying warm strategy of using a lot of layers, and to keep my neck and head covered and warm seemed to do the trick. The second half of the course required us to do a lot of climbing, taking us to 9,500 feet in elevation. Due to the weather, there had been a lot of recent snow up to 2 1/2 feet deep at certain points, so they had to reroute a small section of the course. From Pole Canyon to Oowah Lake which was mile 184.3-201, with an added two miles to the course. Even with the reroute, there was still a lot of snow, and unfortunately, ice. We also had to do 5,400 feet of climbing in this 21-mile section which was very steep at times. I started this section at 6 pm and was out there until 2:30 when I arrived at Oowah Lake. I was fine with the cold, wind, snow, and slow sections of ice. What had me concerned was my water. At about mile 10 of this section, when the temperatures dropped, my water froze solid in my pack. I put a down jacket over me and my pack hoping my body heat would melt the ice, but no such luck. I came into the aid station extremely thirsty and drank cup after cup of apple juice and tailwind. Then after 2 cups of soup, I went to lay down in the car and had a nice 2 1/2 hour nap....snoring and all (Richard also napped some after working this aid station). It was so cold outside and I was so warm in the car that it was VERY difficult to get up and get moving. Only 40 miles left of the race and I was determined to get it knocked out.

I was well on my way on this section of the course when I came across a runner that was moving rather slow. As usual, I stopped to chat with him for a while to make sure he was alright. He was suffering from some severe blisters and said that he was starting to hallucinate. I had a fresh bottle of five-hour energy that I offered to him and he gladly accepted. That seemed to help him. Later on, in this section, I wished I had had some five-hour energy, but was happy to help out a fellow runner, especially if he was hallucinating. I had more at the next aid station when I met up with Richard.

I made it to the very last aid station called "Porcupine Rim", thrilled that I only had 16.5 miles to go to the finish line. I had plenty of time on my side and just tried to enjoy this last section of the trail as much as possible. After all of this time, all the work this year, all the traveling, this was the last little bit of the race....yay and yikes! I soaked in all the feelings and emotions that go along with accomplishing so much in one year. This section of the trail was along a mountain bike course for most of it, so a lot of other people were out and around us. Some even asked what we were all doing out there. One mountain biker stopped riding for a moment to have a small conversation with me and this is how it went:

Biker: What race are you doing?

Me: A 240-mile foot race that started on Friday morning.

Biker: Cool, that's a long stage race.

Me: No, we all are going the whole distance on our own, time doesn't stop. That's why we all look half dead.

Biker: Oh! You look so clean?!

Me: (I thought) that's why I wear dark colors so the stains don't show from the naps I took in the clay when I was so tired, lol.

When I was about a mile away from the finish line I took a small break to sit on a rock along the path. I took a moment to close my eyes and give thanks to all that I have been able to do this year. So much has happened, and I could not be more grateful for how everything played out. I'm incredibly fortunate on so many levels, and I just took a moment for it all to sink in. Then I jogged to the finish line, which made me an official Triple Crown runner of the 200-mile races. I was so thrilled and happy to cross that finish line! Richard, along with all my other fellow runners was there to greet and congratulate me. What an incredible journey! I did it!!

After the race, we took some pictures, had something to eat, and talked with some of the runners before heading to our hotel room. Then I took a much needed hot shower, which felt phenomenal. I chatted with Matt on the phone briefly before going to sleep for the next nine hours. The next day was the award ceremony at 11 am. It was fun to see all the other runners that had completed the Triple Crown and listen to some of the stories they all had to share. So many of us are on different journeys, accomplished so much this year, and it was not only amazing to be a part of it but fun to hear other people's perspectives on what they went through, and what their goals were.

Getting home, back to Charlotte was quite an adventure in itself. We returned the rental car that day at 4 pm and then hung around the airport until our flight at 11 pm. Before boarding we learned the flight was overbooked by 33 people and Delta was asking for volunteers to go on a later flight. I volunteered to go later since my schedule was flexible. Since I volunteered, the airline was very generous, with some prompting by Richard, and put me on a first-class flight, paid for a hotel room, and then gave me a voucher for a good amount of money. I was thrilled. I finally made it home Thursday afternoon and was so happy to see Matt!

What a year of adventure! I lived to tell the tale and am grateful beyond words and humbled by my year on the trails.

I was fortunate to not have too many physical injuries or problems during this race. The worst of them was the blisters on my feet. I don't usually get blisters on my left foot, but this race was different. I ended up losing my big toenail on my left foot and my third toenail as well. From all the races, that left me with 6 toenails but priceless memories, lol. I hope the nails grow back but if they don't, I don't mind, toenail polish goes right on the skin and looks just fine.

My final thought:

Of course, we all have greater inner strength than we can fathom but it does help tremendously to have a solid home base system of support. I urge everyone to tap into your inner strength to accomplish anything you have a passion for. To surround yourself with people that love and support you and that enhance your life because it's up to us to go for it. This year I went for it and am a changed woman because of it. A much more confident, brave, and most of all happy woman, which will not only significantly enhance my life, but everyone else I have the pleasure to have this human experience with. Find your joy and happiness and go out and live it; that's what life is all about. Thank you for coming on this amazing journey with me. Love to everyone :-).

Made in the USA
Middletown, DE
29 May 2019